BOAT MAINTENANCE GUIDES

The
RESTORATION
Handbook For Yachts

Dedicated to my grandfather, Don Fernando Rubió i Tudurí,
who taught me to love and respect the sea

BOAT MAINTENANCE GUIDES

The
RESTORATION
Handbook For Yachts

ENRIC ROSELLÓ

THE ESSENTIAL GUIDE TO FIBREGLASS
YACHT RESTORATION & REPAIR

FERNHURST
BOOKS

Reprinted in 2018, 2020 & 2023 by Fernhurst Books Limited

Second edition published in 2018 by Fernhurst Books Limited
The Windmill, Mill Lane, Harbury, Leamington Spa, Warwickshire. CV33 9HP. UK
Tel: +44 (0) 1926 337488 | www.fernhurstbooks.com

A catalogue record for this book is available from the British Library
ISBN 978-1-912177-13-4

Front cover photograph courtesy of Alfred Farre

Typeset in India by Laserwords Private Limited, Chennai
Printed in the UK by CPI

Table of Contents

Foreword

It was, I believe, Sir Francis Chichester who said that one should never buy a yacht unless it was a case of love at first sight. And this is exactly what happened to me the first time I set eyes on *Samba*, moored at Bayona on a grey winter's morning, after having arranged to take her for a short trial run. I had no doubt that her previous owner must have felt the same way about this marvellous Angus Primrose design, with her low freeboard and raked stem, an ample beam promising firm stability and her fine narrow stern, culminating in a classical triangular transom, long since replaced by today's designers to allow for more extensive bathing platforms.

I realised what I could do with this boat as soon as I opened the door to her stern locker, containing the canvas, cordage, tools, waterproofs, her outboard… there were even a couple of long oars. This was a boat that had sailed a lot and done it well, and her owner had evidently treated her with the affection that only seafaring people can truly feel for their boats.

Sailing her off Bayona, on a rolling sea, I soon discovered the qualities of that design of which, if my memory serves me well, *Samba* was only the second to be built in Spain. Later, during a first voyage to Sardinia with an inexperienced crew – and this was over twenty years ago – battered by a ferocious mistral that had forced me to use the storm jib and put a third reef into the mainsail, *Samba* made it more than clear that I could trust her fully, in any seas. That was the first thing she taught me and something that I never doubted afterwards. Ramón Massachs, who first showed me how to tack, warned me as he handed her over in Almeria, after bringing her through the straights in the face of a fierce easterly, "This boat will always be more capable than you". And so she was for all the years that I sailed her.

Perhaps that was why, when I eventually sold her, I was filled with an enormous sense of emptiness. It was neither because I was disappointed in her nor to replace her with a more spacious or modern boat. I felt absurdly guilty, as if I had turned my back on an old friend, for no good reason, I could not justify it even to myself. But life is full of surprises and you never know what the future will bring.

Years later I got a call from my colleague, Enric Roselló, who had bought *Samba* and told me of his plans to refit her. Shortly afterwards I visited the slipway, down at Arenys de Mar, and climbed up the ladder that was laid against her side, and what a sorry sight she was.

Right there and then all of the sensations that I had had the first time I boarded her in Bayona came flooding back. Even with her interiors completely stripped out for the refit *Samba* still gave off that same, unmistakable smell of salt and diesel. The picture of the Virgen del Carmen, placed there by her first owner, and which for many years nobody had dared to remove from its discreet position on the bulkhead, was no longer there – tradition and superstition have always accompanied seafaring people. Enric had not been able to find the peseta coin, which her original builders had glued to the bottom of the well to bring good luck, although she had evidently had her share, because there she was, alive and well, and now in the best possible hands, those of a friend.

It's not easy for an owner to get hold of a classic fibreglass boat and give her a thorough refit and renovation. "There's only one reason for doing it" I thought "It must have been a case of love at first sight". And I left *Samba* satisfied and fully convinced that this boat, which used to be 'my boat', had rightfully earned herself a long, second life.

Germán de Soler

Author's note: The small portrait of the Virgen del Carmen and the peseta coin were put back in their places when the restoration work was finished.

Preface

Even after *Samba*'s refitting work was finished I still had my doubts, right up to the last minute, about the best way to organise this book, in terms of the sequence of the work that was done on her. On the one hand it seemed logical and comprehensible to group the work thematically. But in reality, like it or not, a complete refit never works out like that. The boat had been rebuilt by a constant influx of alternating specialists. In the end I decided, as far as possible, to respect the chronological order in which the work had been done, even in the knowledge that the list of contents might seem chaotic to the reader.

I cannot finish this short preface without reiterating my gratitude to the different nautical professionals and tradesmen that contributed to *Samba*'s refit, and who were subjected to additional stress by the continuous photographic sessions and a non-stop barrage of technical questions that I would need to know the answers to in order to write this book.

I must also thank the managers in charge of the companies, whose material we used during the refit, for their support and opinions. Thanks to all of them, this refit did not end in the kind of disaster that is so often the result of the best and most innocent intentions of meddling amateurs such as myself.

Enric Roselló

Before starting to work on a boat it is a good idea to make a list of the problems that need to be solved and have as clear an idea as possible of the overall approach of the refit.

In search of a second youth

Making an inventory of the jobs that lie ahead is much more important than you may think. From this first inventory must emerge the first ideas, approximate as they may be, regarding the budget for the repair work, how long it is going to take and the necessary planning of the different stages.

Few yacht owners have either the technical know-how or enough free time to completely rebuild or restore a boat themselves. The usual situation for boat owners is that they find themselves with a boat, either a second-hand boat or one that used to be new but no longer is, and are prepared to carry out, on their own account, some kind of limited repair or refitting work.

There is no clear dividing line that marks the level of technical difficulty that amateurs should not cross in terms of the work that needs to be done on their boats. And in the same way that the best professional mechanic is not necessarily a great painter, some amateurs will find their métier replacing the electrical wiring and, at the same time, will be incapable of cutting a plank of wood with a handsaw. Or vice versa.

Honestly appraising your own do-it-yourself ability, with all of its pros and cons, forms a vital part of the initial refit inventory. From this process of self-evaluation a series of jobs will emerge that you feel you can take on yourself, others for which you will unavoidably need professional help and some that you will be able to do yourself as long as you can pick up a few tips. We hope that this book will help you with this breakdown.

A complete refit

We were going to renovate *Samba* from keel to masthead including all the eqipment between, and her interiors. The solutions that we would be adopting were not absolutes, it could be possible to do them better, worse or simply in a different way altogether, it is not a question of laying down the law. Our intention here is to contribute ideas and working methods that might prove to be of help to anyone finding themselves in similar positions to the ones we were in.

By presenting the different jobs in independent chapters we are also offering the opportunity to limit yourself to specific areas. One year you might be interested in replacing the water installation or the electrical wiring, while the next you may wish to subject the hull to treatment against osmosis or to strip her out and redo the interior furnishing.

For our part we will try to explain each of the jobs that we took on in a practical sense, explaining what we did and why and giving you some idea of how long it took us. This will be accompanied by a sequence of photos that illustrate these different processes. In this way we hope that some of you may pluck up the courage to do the work yourselves while those that decide to entrust the work to professionals will have a good idea of what is involved and what bills they can expect.

Thirty years before the mast

Samba was originally delivered to her first owner, a charter company, in kit form and the initial layout of her interior provided space for as many as 11 bunks. After a short interval as a privately owned boat she was sold again, returning to charter use in the eighties and early nineties. During these years some of her berths were removed and her accommodation space summarily refitted.

For one reason or another, she never managed to settle on a ultimate interior finish, which little by little began to lose its shine and comfort. Much the same thing was also happening up on deck where, since the day she was launched, the only work that had ever been done was limited to fixing whatever got broken.

Our aim was to rescue *Samba* from her ignominy and to return her dignity to her. A hull as secure, good looking and seamanlike deserves nothing less. There is one consideration that we feel it is important to make at this point: one of the first mistakes that an amateur must avoid when faced with an important

refit is choosing the wrong boat. When considering the possibility of investing a lot of time and money on rebuilding your boat you have to be sure that the model of the boat in question, and her sailing characteristics, are as close as possible to what you are actually looking for. If this is not the case it is far too easy to become discouraged at any time during the refit, or even after you have finished the work.

Metres
Feet   SCALE 1:24

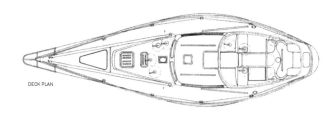

DECK PLAN

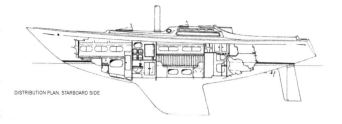

DISTRIBUTION PLAN, STARBOARD SIDE

North Wind 40:
A real classic

The North Wind 40, designed by a British man, Angus Primrose, was very active for almost two decades in many international and local races. This was the boat in which José Luis Ugarte finished his first and hard single-handed Transat. Many years later, the veteran North Wind 40 *Orion Iru* continued to be this great Basque sailor's own personal yacht.

Angus Primrose's fame was endorsed by the racing success of many of his designs, including Sir Francis Chichester's *Gipsy Moth IV*. Primrose also had a long-term collaboration with Moody, where he designed the majority of this boatyard's models in the seventies. After Primrose's premature death, Bill Dixon, his chief assistant at the time, took over the office and has continued to collaborate with Moody up to the present time.

The North Wind 40 story started in November 1972, when Primrose drew up the plans for a prototype that, sailed by Bruce Banks, finished second in the trials for the British Admiral's Cup Team. Barcelona-based company Manufacturas Mistral bought the plans for this yacht and managed to build the first 12 boats before the boatyard closed in 1975. Industrial North Wind (nothing to do with the current company of the same name) then acquired the mould of the boat one year later and went on to build another 25 boats.

In 1980 the original model was redesigned, to become the North Wind 38, with her stern trimmed, a new keel and the accommodation laid out with more cabins. At least 35 examples of this new version were built before the final closure of the boatyard in 1984.

The success of the North Wind 40 was based on her good performance and sturdiness. Her silhouette, with low freeboard, notable beam and fine sharp stern are the marks of identity of the years in which she was designed. Yet even now, in the 21st century, this slim and classical hull shape still evokes the spontaneous admiration of anyone who loves a yacht with personality.

The North Wind 40's trapezoidal keel (2-metre draught) ends in a slightly bulbous form. Including the keel, over 50% of the boat's weight is below the waterline, which explains her hull's excellent stability when heeling, the way she cuts through the waves and her capacity to fetch to windward.

The section of the mast, rigged at the masthead with spreaders, is generous and, for the majority of boats in service, has withstood the test of time. The same could also be said about the reinforcement bulkheads and laminate bulkheads, which have rarely showed signs of weakening.

Leaving apart questions such as finish or maintenance, which certainly must vary from one boat to another, where the North Wind 40 has suffered most from the passage of time is in the concept of her manageability which, despite being complete, has become rather obsolete. The winches on the mast or at its base and her antiquated returns are not very effective in terms of the possibilities offered by modern-day fittings.

The North Wind 40's accommodation also suffers from a lack of volume aft (low ceilings were prevalent in earlier decades) and has only two cabins for a boat with a length of almost 40 feet. On the other hand, the ample stowage capacity that she can offer is equal to, if not greater than, the space most modern yachts have to offer.

Specification sheet
Model: North Wind 40, **Designer**: Angus Primrose, **Builder**: Manufacturas Mistral y Industrial North Wind, **Years built**: 1973 - 1984, **Building materials**: fibreglass, **Overall length**: 12.79 m, **Hull length**: 12 m, **Waterline length**: 9.16 m, **Beam**: 3.72 m, **Draft**: 1.98 m, **Displacement**: 8,450 kg, **Ballast**: 4,070 kg, **Engine**: 25/50 HP., **Freshwater**: 600 litres, **Diesel**: 300 litres, **Mailsail**: 24.6 m², **Genoa**: 60.4 m², **Spinnaker**: 140 m²

Step by step

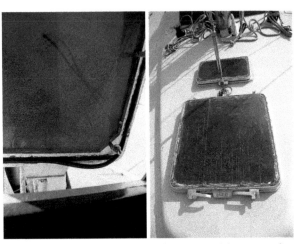

➤ Hatches and companions. As you would expect after almost thirty years of exposure to the elements, the Plexi-glass of her front windows had become completely glazed, to the point where they were almost opaque. These had to be replaced, as did the boat's different hatches, including weather seals and badly worn fittings.

➤ Painting the hull and deck. When it comes to fixing up a boat there are a few jobs that work as effectively as a good coat of paint on hull and deck. Samba's gelcoat, thoroughly weathered, lacking shine and with the wear of twenty years of service was just crying out for this to be done.

➤ Redesigning the rigging. Although the North Wind 40's rigging is complete and relatively straightforward, its antiquated layout, around the base of the mast, was a nuisance. If we wanted to be able to manoeuvre the boat with a reduced crew we would have to modernise it by running the halyard returns right into the cockpit.

➤ Interior furnishing. Originally delivered in kit form, Samba was never furnished to the level that she really deserved. A number of changes made to the layout of her accommodation over the years, always done amateurishly and without any great care being taken have, along with pretty slipshod maintenance, left a hotchpotch of mismatched finishes that were sadly in need of sorting out.

➤ **Galley.** While conserving the useful size and overall layout of this galley, both the furnishings and equipment needed a complete overhaul if they were going to be used again.

➤ **Head and shower.** The actual head was truly austere, not to say minimalist, despite the pressure unit and shower that were installed a few years ago. On the port side you could still see the remains of the original berths, now converted into improvised shelves. An in-depth refit would be very welcome indeed.

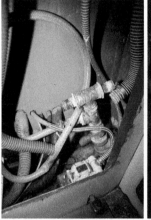

➤ **Water and drainage fittings.** The water system, obsolete in all sections, would have to be completely replaced. Another of Samba's dark areas was the drainage system, which had also earned retirement for its component parts. This was a question that would need serious consideration.

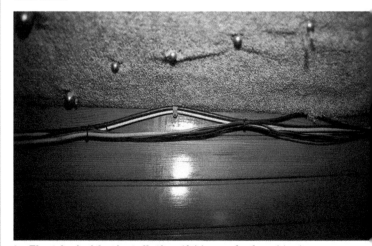

➤ **Forward cabin.** The forward cabin was relatively large for Samba's overall length. If nothing else we would need to fit some decent stowage space, as well as giving the mismatched finishes a thorough once over.

➤ **Electrical wiring installation.** If this was far from ideal in its day, after over thirty years of service the whole of the electrical wiring installation was in urgent need of replacement, in order to make the boat safer and more prepared for the demands of modern life.

➤ **Electronics.** *The boat's electronics were also long past their retirement date. The advances in this sector over recent years would make any attempt at repairing the existing equipment a waste of time.*

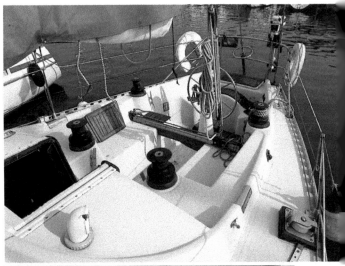

➤ **Upholstery and foam rubber.** *Seating and berths are an important element of comfort onboard. Thoroughly worn out through use, Samba's old upholstery and foam rubbers would have to be replaced.*

➤ **Deck works and fittings.** *The deck fittings were a disaster. Except for the cockpit winches, which were in a reasonable condition, all the rest was old, broken down or jammed (if not all three at once). A complete update was essential.*

➤ **Linings and claddings.** *The claddings on the ceiling and interior textile linings were another of the boat's aesthetic aspects that we were going to be taking a close look at.*

➤ **Engine.** *Unusually robust and reliable, her original 40 HP Perkins 4.108 was now well past its useful work-ing life. We had to consider whether it would be worth repairing or whether it would be simpler just to go ahead and replace it.*

Every refit tends to start with the boat hanging from the straps of the travel lift. When faced with long and exhaustive work on hull and cabins, bringing her on dry land is essential.

High and dry, stripped out and inventory

Taking *Samba* to the slipway with the idea of leaving her in dry dock for some time was the first step towards her renovation. The first thing we did then was strip her out completely, starting with her lockers and stowage spaces. Emptying out the contents of your boat's lockers, chests, drawers and stowage spaces is something that all owners should do at regular intervals. It is astonishing how much junk and clutter, most of it of no use whatsoever, will accumulate in a boat over the years.

Removing everything, classifying it, arranging it and then returning only what is necessary, is fundamental for maintaining some kind of order and gaining space on board, which is always valuable. There is also the healthy custom of keeping unnecessary weight, which will always restrict your boat's performance, under tight control.

In the case of *Samba*, once on dry land, first we threw away everything that was obviously broken or served no purpose and then we organised and classified the myriad of accessories, sails and stores, packing them in boxes and bags and carting them off to be stored on land, in dry storage, where they would have to remain until the work on the boat had been completed.

A first recommendation for anyone working on a refit is not to throw anything away (unless obviously broken and irreparable) until it has been satisfactorily replaced. In our case, for example, even though we knew that a good many of the sails, running rigging, life jackets and other stuff were crying out to be retired, we gave them a thorough cleaning with soap and water and then folded them up neatly and stored them away. Should things start to go wrong, or should the budget fall short of covering that winder, a new genoa or a brand new set of top quality ropes, then we would always have the old ones to keep us going until better days came round.

Working in dry dock

For repairs affecting different technical areas it is important that the chosen shipyard offers all of the services that you are going to require, or can at least subcontract them out to competent professionals, or else you will end up having to take the boat to different places for every little job. If in doubt, ask for references about other boats that have had similar work done. No serious professional would deny this request.

Practically all shipyards will allow boat owners to do such repair work as they consider expedient themselves. But if you have a mechanic, a painter or a carpenter that you have confidence in, ask whether there would be any problem with them working in the shipyard that you have chosen, just in case such work is covered exclusively by another professional. In short, in order to avoid misunderstandings in the future, it is far better to clarify these questions before you start.

If the owner is going to do some or all of the work on the boat by himself, it is also important to find out whether he will be able to do it at weekends and/or during holiday periods (like Christmas, Easter, bank holidays, etc.).

➤ *The first step in the restoration of* Samba *was to bring her up to the dry dock.*

Inland restoration

An option often chosen for long amateur refits is to take the boat away from the sea. This is often done by people who have access to a piece of land, an industrial building or any kind of covered space where they can leave the boat close to where they live. This solution has its advantages, such as saving on yard expenses and, perhaps more important still, the accessibility of this solution for people who do not actually live near the coast.

The experience and know-how of nautical tradesmen is irreplaceable, although this does not mean that any carpenter, car bodywork painter or mechanic specialising in tractors cannot provide help or advice for many of the jobs that need to be done on board. The main differences in jobs on board, rather than working methods as such, are the materials that you have to use. A domestic door is similar, probably more complex, than its nautical equivalent. But on a boat, specific plywoods, woods and glues have to be used. The same can also be said about painting, electrics or mechanics.

Step by step

➤ *Most of the restorations start with the boat suspended from the straps of the travel lift.*

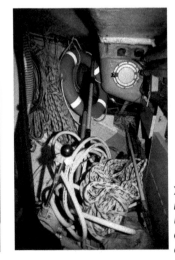

➤ *It is incredible how much junk, most of it useless, will accumulate over time in the lockers of any boat.*

➤ Completely emptying out the boat is something that all boat owners should do every now and then.

➤ Sails were thoroughly washed with fresh water and soap before being dried out and folded up. We did the same with the lifejackets, running rigging, gear and any other accessories that might have been in contact with salt water.

➤ Removing everything, classifying it, putting it in order and returning only what is needed is a fundamental part of keeping on top of things and gaining space on board, always precious, not to mention getting rid of all that useless weight.

➤ In order to pave the way for work in the future, instead of returning everything that might be useful to the boat, we took it to a dry and protected place, on land, for long term storage while the boat was in dry dock.

➤ During a refit nothing should be thrown away until it has been replaced. In our case, even though we knew that the sails and running rigging were begging to be retired, we kept them clean, dry and neatly folded. If things started to get difficult, and should the final budget not cover that headsail furler, new genoa or top quality ropes, then we would always have the old ones to fall back on until we could afford new ones.

Chapter 3

Stripping out interiors

The usual approach when starting a refit is to strip down until you can find a secure base on which to start to build up again. This allows you to get to know the terrain and have an idea of the requirements, defects, strengths and weaknesses of each possible solution.

The good news and the bad

Little by little, behind panelling, inside lockers and under linings, different coats of paint started to appear, a full range of nuts and bolts, screws, electrical wiring and water pipes that run from nowhere to nowhere, or through-hulls that have served no purpose for a long time. *Samba* was originally delivered with a number of her bulkheads painted and a fine wood veneer glued to the most visible wall surfaces of the saloon, and the chart table.

Successive owners then went about adapting the layout and the decoration to their own taste, although this was done in a rather rough and ready way and without fully covering over the traces of everything that was being replaced. The old through-hulls and sea cocks, from where the original head had been (at the bow) were, for example, still in place, blocked with corks for safety (sic). This is a time-bomb in terms of the water tightness of any boat!

The boat's last owner, with good intentions but lacking means, covered over most of the interior panels with endless small wooden chocks, each joined to the others in a rudimentary kind of tongue and groove system. Sometimes with screws, often glued, the aesthetic aspect of these chocks left a great deal to be desired and stripping them out, one by one, was laborious work.

Although stripping down can be heavy work it does not require a great deal of technical know-how or skill. An amateur with average skills can manage this on his own, while it also offers an ideal opportunity to gather first-hand knowledge of your boat's hidden weaknesses and allows you to confront the refit in full awareness of some of the problems you are likely to come up against.

Finally, I would also like to add that, while it may be tempting to think about saving a few quid you should never reuse the endless screws that you will find when stripping the boat out. The advantages of such a saving will never come close to compensating for the endless headaches that will result from using screws that may have damaged heads, bases or threads, even if they look OK.

Getting down to basics

The decorative panels, also tongue and groove, in the saloon and bathroom were screwed to a plywood panel that was, in turn, fixed to the hull using iron bolts. The accumulated rust on the iron bolts turned out to be a nightmare when it came to removing them. How easy it would have been to use stainless steel bolts!

We discovered most of the electrical wiring, hanging loose behind the panels with no channels or protection of any kind. It would have been impossible to check or change a single wire without first removing these panels. Clearing out the boat makes the replacement of water pipes and electrical wiring so much easier, rather than trying to do it with all the furniture, cladding and linings fully assembled and in place.

A pleasant surprise, when we got down to the boat's entrails, was to discover that there were hardly any problems with her bulkheads and structural reinforcements, in terms of rot or delamination. Discovering such surprises, both pleasant and otherwise, is another of the rewards of stripping your boat down to her basics.

With regard to the fine wood veneer of the bulkheads, this is a cheap solution typical of boats that were delivered in kit form. In the photos it may look like this veneer was in good condition but it had become partially unglued and was swollen and bulging due to the effects of the damp and ultraviolet light. Perhaps this was an acceptable finish in its time but these days it does not do the boat as a whole any favours. After four hours of absorbing work with the electric scraper all of the veneer had been removed. Later on we will find a replacement for it.

With practically all of her fibreglass exposed, *Samba*'s interiors had taken on the lean aspect of a racing yacht. From this point on we could start to think about rebuilding her interiors.

Step by step

Stripping down: before and after

Saloon

Forward cabin

Head

Galley

➤ This pile of wood and carpet scraps, 'dumped' over-board, was the result of a single morning's work. To strip down Samba's accommodation and leave it ready for building work to start took us forty hours of work.

➤ The decorative finish of a number of the interior bulk-heads consisted of a wide range of different wooden chocks and strips, all joined together in a curious collage.

➤ Sometimes held in place using screws and sometimes glued, these wooden strips provided an aesthetic aspect that left a great deal to be desired. Stripping them out, one by one, was laborious work.

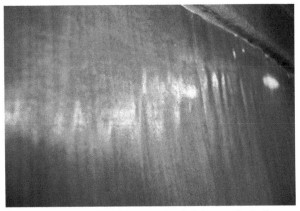

➤ Although the bulkhead veneer finish may look decent from a distance, close up you can see how it has become partially unglued and/or is swollen and bulging as a result of the damp and ultraviolet light.

➤ After four hours of absorbing work with the electric scraper we managed to get all the veneer off.

➤ Most of the electrical wiring was hidden away behind the decorative panels in the saloon, without channelling or protection of any kind. It would have been impossible to check or change a single one of the boat's wires without removing these panels. With practically all the fibreglass exposed, at this point Samba's interiors had taken on the stripped down aspect of an authentic racing boat.

Through-hulls: a latent danger

It is impossible to be dogmatic about the useful life of a through-hull despite their vital role in a boat's safety. It all depends on the care that has been taken of them, the level of damp in the area where they are fitted, the quality of the components used etc. When in doubt the best thing is to change them every five years. They are not particularly expensive and are quite easy to replace. However, in the case of defects, they can ruin life aboard and could even sink the boat in a matter of minutes.

A seacock needs very little care to work correctly. Spraying it with a lubricant every four to six months is sufficient to guarantee its useful life. During (at least) each application a check ought to be run to see that they are working correctly. You must be able to turn the handles smoothly and the valves must be completely open or closed, depending on the position of the handle. Should the handle resist movement it must never be forced (above all with the boat in the water). The bronze of the through-hull is a relatively soft metal and may break without warning. But, most important of all, if corrosion or electrolysis are on the march, never wait until the last moment before replacing them.

Through-hulls and their seacocks ought to be connected to an earth to avoid electrolysis, as is the case with all of the boat's metal parts that are in contact with the water. This is a detail that you will actually find on very few boats, which means that you have to pay particular attention to this kind of corrosion, which acts from the inside and does not become visible until it is too late.

➤ A through-hull in bad condition can ruin life on board and may even sink the boat in just a few minutes.

➤ *Through-hulls and seacocks from Samba's original heads (located at the bow end) were still in place twenty years after they were first fitted. A simple bung had been used to block the PVC valve, a material to be avoided in nautical plumbing.*

Removing a recalcitrant through-hull

Between electrolysis, corrosion and even silicone, almost all of *Samba*'s old through-hulls were seized. The problem becomes worse when, as is often the case, there is hardly any space between furnishings to use tools with any great leverage capacity. The solution is to work on the through-hull from the outside using a circular electric grinder or a cold chisel, depending on the size of the through-hull and its consistency.

➤ *Sometimes it is impossible to remove a seized through-hull from inside the boat, as you cannot get at it either with tools that have some leverage capacity or with the electric grinder. Over time electrolysis and corrosion become a harder sealant than the most powerful glues.*

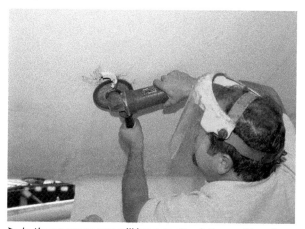

➤ *In these cases you will have to attack the problem from the outside, using an electric grinder and/or a cold chisel. Due to the abrasive power of the grinder you must take utmost care to avoid marking the fibreglass. We would also remind you that eye protection is essential when working with either metals or fibreglass.*

➤ In some cases a couple of blows with a cold chisel are enough to break off the base of the through-hull while in others you will need to cut the base into segments like a cake using a grinder and later finish off the work (using either the grinder or a chisel) segment by segment.

➤ The base will start to give quite rapidly and in about twenty minutes the most recalcitrant through-hull can be removed.

Removing the mast: a complete overhaul is imperative

An annual check on the mast and rigging, just visual to begin with, is always recommended. Climbing up to the masthead to check blocks and their pad eyes, wind vane, aerials and navigation lights, and then coming down, little by little, checking all of the shroud anchors on the mast and the spreaders, and finally checking the bottle screws and chainplates. These are all jobs that owners should either do themselves or entrust to a competent professional before the start of each season. Not doing so could lead to serious problems.

It does not take long to remove a mast, although it does represent a significant extra cost in your budget. In view of the meticulous inspection we were making of the boat, after such a long time, the decision seemed more than justified.

➤ To remove the mast in an easier and more convenient way all the halyards were coiled up at the foot and securely tied off using duct tape. Once the crane had a firm hold on the mast the last four shrouds holding it in place were also loosened and then tied off to the mast, using shockcord.

➤ This avoids them hanging loose and getting in the way as the crane is lowering the mast to the ground.

➤ We do not know whether Samba's mast had been removed at any time during her thirty years of service. By the look of the accumulated dirt that had built up inside the mast partner hole it seems quite possible that this was the first time the mast had been removed since she was launched. An in-depth examination would be called for.

Chapter 4
Osmosis

Over the years osmosis will always appear, in one form or another. In this chapter we will attempt to explain this inevitable 'disease' of fibreglass, as well as how it can be treated.

An inevitable problem

In this book there will be two different, and widely separated, chapters dealing with the diagnosis and repair of osmosis, and providing a true reflection of the reality, in which several months must be allowed to pass between peeling the boat's bottom and repairing it. In the case of major refits it is a good idea to take the first opportunity to approach the problem of osmosis, to start drying out the hull, while you get on with work in other areas.

We were hoping that the osmosis on *Samba*'s hull was not too serious. An epoxy treatment that she was given a number of years ago gave us some grounds for hope. After blasting a few areas, to expose the gelcoat, the next operation was to measure the level of damp at these points. Our hopes were short-lived: we found rampant osmosis.

In the red

During the verification process applied to sample areas of the hull the indicating needle barely managed to get out of the red, for some measurements going right off the scale. It was going to be essential to thoroughly cleanse everything and allow the fibreglass to dry out.

Proprietory moisture meters are not fitted with a decimal scale of any kind, what they actually do is quantify the electrical conductivity (remember water is a good conductor of electricity) between two points. In boats with fibreglass bottoms, and this is also true for wooden-bottomed boats, the greater the conductivity the more water has been absorbed by the hull. On the other hand, if we apply this meter to a metal the needle will also indicate the highest level even though there is no water. The indicator on these meters is like a traffic light; there is a green area (little or no humidity), an amber area (some humidity) and a red area (where humidity is higher than the recommended limits).

A boat that is afloat has all the numbers in the osmosis lottery, insofar as osmosis will arrive sooner or later and to a lesser or greater extent. Logically, therefore, a boat that winters out of the water will have a much better chance of avoiding osmosis unless, of course, it has stagnant water in its bilges, a possibility that many are ignorant of but which also causes osmosis.

Over time the humidity spreads through the different fibreglass laminates although, luckily, it never rises vertically above the waterline. This is why osmosis only ever affects the part of the hull that is below this line.

So what is osmosis?

Osmosis is a degenerative process that occurs inside the fibreglass laminates when water has penetrated through the gelcoat to them. Despite its excellent properties, gelcoat (derived from polyester) is neither one hundred per cent waterproof nor waterproof throughout the useful life of a boat. Little by little over the years it loses its impermeability (cracks, pores, knocks, natural wear and tear, etc.) and progressively allows water to seep through into the fibreglass.

Once the gelcoat layer has been breached then seawater will start to react chemically with the polyester leading to the decomposition of the polyester and the production of acetic acid. This substance, given that it is denser than water, cannot pass back through the gelcoat to the outside and will build up between the fibreglass laminates, exerting pressure on them and forming blisters in the 'skin' of the hull. When these blisters burst the liquid contained in them can be easily distinguished by its greasy feel and vinegary smell.

The famous blisters

So, the symptoms of osmosis are these blisters, which discharge acid through the gelcoat and many an owner of a boat of a certain age will smile in relief thinking that her hull has miraculously escaped osmosis because they have not found any blisters. Do not be fooled, the blisters are only a symptom; the disease is the penetration of water into the fibreglass laminates. This is not visible to the naked eye and requires the use of a moisture detector, the supreme and only judge in this type of investigation.

In relatively new boats (up to approx. ten years of age) the gelcoat continues to partially play its protective role. On the one hand it allows water to penetrate into the laminates but still prevents the denser osmotic fluid from passing the other way, giving rise to the famous acetic acid blisters.

In the case of older boats, such as *Samba*, gener-

ally speaking their hulls do not become blistered. What happens in these cases is that the gelcoat becomes so permeable that it no longer provides a barrier to either salt water coming in or acetic acid going out. This means that the osmotic fluid will escape through the gelcoat and as a result does not form blisters. In either case the problem is the same, osmosis delaminates the fibreglass through the decomposition of the polyester that binds it.

In all cases the solution lies in recreating the protective barrier between the fibreglass and the seawater. What is usually done, rather than reconstituting the gelcoat, is to apply a coat of resin and/or epoxy filler, a more expensive protection but one that is much harder and more impermeable. But first of all the hull has to be completely dried out, because if not the accumulated water will continue to do its work and the blisters will appear again a few months after the treatment.

This is why preventive treatments (generally consisting of coats of epoxy resin applied directly to the gelcoat) only work on new boats or boats that have been in the water for at most six months. If they have been in the water longer, their effectiveness is highly debatable. In cases of incipient osmosis it is better to wait one or two seasons and then give the boat the full treatment rather than a merely preventive one.

There is a solution for everything

A few years ago, when it became obvious that fibreglass hulls, which had previously been thought to be eternal, suffered from osmosis, some reckoned that would be the end of this material in the construction of hulls.

Fortunately modern treatments with an epoxy base are a perfect and relatively simple solution for this inevitable fibreglass problem. Once osmosis has been

➤ *Blasting with water and sand selectively cleanses the blisters and the worst patches of fibreglass damaged by osmosis.*

detected all you have to do is get a price and leave it in the hands of a good professional tradesman (amateurs have little to offer in this area) who can return her as good as new within a few months, ready to face another 15-20 years at sea without problems.

Paradoxically, in a 40-footer it is more expensive to replace her canvas (main, genoa and spinnaker) than solve the problem of osmosis. As a result it is surprising that many buyers only check the level of moisture in the hull when they buy a used boat, without even checking the state of her canvas, engine or other parts of the boat that may have defects that would cost much more to put right than osmosis.

Evidently osmosis is cause for alarm for any owner but it would take a catastrophic lack of care for this problem to actually weaken the structure of the boat, through the process of osmotic delamination. A great deal of margin exists before the point of no return is reached.

A good clean

To eliminate the gelcoat, the first stage in any osmosis treatment, the best solution is to use an electrical scraper followed by blasting of the most damaged patches of fibreglass using sand and water. If the gelcoat is not eliminated it could take years for the hull to dry out. Scraping is done using an adjustable electrically-powered wire brush, which will eliminate the gelcoat without affecting the fibreglass.

For healthy boats, blasting the hull with sand and water under pressure is used to thoroughly eliminate the coats of paint, exposing the gelcoat. When used

after the wire-brushing of the gelcoat the blasting selectively removes the blisters and the fibreglass laminates damaged by the osmosis. Experience is required for both wire-brushing and blasting, which means that it is work that must be done by professional tradesmen experienced in treating the bottoms of boats.

In terms of how long it takes for a hull to dry out, this is hard to say. It really depends on the level of moisture contained, the type of resin and fibreglass that the hull is made of, the ambient temperature, level of humidity, etc. Generally speaking three to six months is enough time for the hull to cure, although there have been cases where over a year was needed to return a hull to acceptable levels of moisture content.

While, at first sight, it may appear that the five to ten litres of water that the hull of a 40-footer could have absorbed due to osmosis ought to dry out much faster, you also have to set about eliminating the salt and acids that have formed between the different laminates. That is why the application of heat, much used in pioneering osmosis treatments, barely affects drying times. Periodic cleaning of the hull, simply hosing it down with fresh water under pressure, is much more effective. This unblocks surface pores and encourages the internal drying of the fibreglass.

Step by step

➤ *Drying time for a hull suffering from osmosis will depend on the level of moisture it contains, the type of resin and fibreglass used, ambient temperature and humidity, etc. Generally speaking three to six months is enough time to cure a hull, although there have been cases where over a year was needed to return one to acceptable levels of moisture content. So we must wait and, in the meantime, there are plenty of other jobs to be getting on with. Samba will dry patiently, like a good 'Serrano' cured ham.*

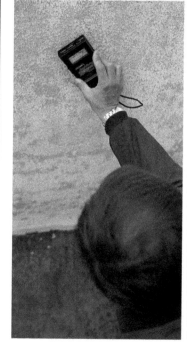

➤ *In older boats, such as Samba, the hull will not usually show signs of blistering; although this does not mean she is free of osmosis (in fact the opposite is the case). Osmosis is not always visible to the naked eye, thus the importance of the moisture meter.*

➤ Blasting also serves to eliminate the coats of paint and prepare the keel for protective treatment. Although it may not appear to be a porous material, cast iron can absorb water up to 10 per cent of its volume (not to be mistaken with its weight). It is a good idea to let it dry out before applying the corresponding impermeability treatments. This will avoid, as far as possible, the formation of oxide.

➤ Elimination of successive layers of paint will expose details such as the epoxy filler seal, applied in a block between the keel and the hull, to avoid (successfully it seems) the entry of water through this part of the boat.

➤ Gelcoat has become so degraded over the years that a simple jet of sand and water under pressure is almost enough to lift it off completely. The numerous craters in the fibreglass will require the meticulous application of layers of resin and epoxy filler which will later have to be sanded down again.

➤ Cleaning isolated areas of the hull will also allow for the state of the boat's various war wounds, gained over years at sea, to be checked, such as this repair to the rudder blade.

➤ Throughout the examination the needle of the moisture meter was barely out of the red, and when measuring some parts of the hull even went off the scale completely.

➤ Stagnant water in bilges, over looked by many, is as much to blame for osmosis as the water that penetrates from the outside.

Some people may be surprised that we were starting our refit of Samba with the electrical installation but there is nothing strange about this: it is an approach that we copied from production boat builders, consummate experts in the optimisation of work methods.

Planning the new installation

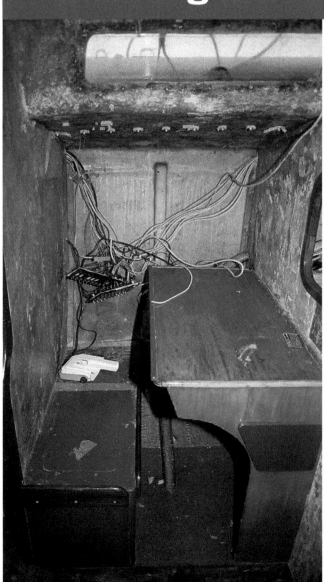

The perfect time for an electrical refit is when the boat is stripped bare so that new wiring can be hidden behind interior panelling. Electrical installation should progress in line with other work but not completed until almost the end of the refit, when the last light sockets are finally connected up.

Electrical installation is one aspect of modern boats that has changed most drastically over the last forty years. Old requirements, including legal ones, have completely changed. Before embarking on any electrical work, you must ensure all safety regulations are met and that the work meets your country's standards for electrical installations on boats. These regulations may be updated without notice (which is why we haven't gone into detail here) with different countries differing in their requirements. A good starting point, would be to contact your country's national governing body for boating.

Modern navigation's electrical needs have also moved on. It is impossible to contemplate modern pleasure sailing without the full deployment of electrical and electronic gadgetry. Nowadays we bring just about every home comfort on board with us, nobody is prepared to go out sailing without a microwave oven, a computer, water treatment, fridge or even a hairdryer.

Accommodating all these home comforts on an old boat requires a great deal of thought. Most electrical installations from thirty years ago were rather slapdash and trying to adapt old circuits to modern demands raises problems that may be difficult to resolve.

Step by step

➤ The above is a small part of Samba's old electrical installation, the wiring distributed around the boat with hardly any protection at all. For the new installation we would be using ducts to run this wiring, updating and improving the layout using a modern technical approach.

➤ There are few things that let a boat down as badly as exposed electrical or electronic wiring running across ceilings or walls.

➤ When we stripped out the interior of the boat we discovered endless wires, stuffed away under fitted carpets and in some of the most unlikely corners, many of which seemed to serve no purpose whatsoever.

➤ The veritable rat's nest of wires connected to the switchboard made it impossible to figure out what was what or to even consider making improvements to the existing installation. In these cases the easiest thing to do is just rip it all out and start again, with new wires and working from new circuit wiring diagrams.

➤ The boat's lighting, and most of the other equipment too, were connected up using a series of dodgy and dangerous looking splices. Bad connections are the cause of the majority of electrical failures on boats.

➤ Behind the switchboard, on modern production boats everything is clean and neat. A DIN strip connects the positive switchboard wires to the boat's installation, while the negative wires are brought together in a single connector. An up-to-date electrical installation was one of our main goals.

Renovating the electrical installation: Plan of action

A complete renovation of the electrical installation requires a level of technical know-how that is normally only available to electricians. Amateurs, duly equipped with installation manuals and/or wiring diagrams to help them with the work, may be able to manage when patching up the old system or carrying out occasional repairs.

Another solution within the reach of amateurs is simply replacing the boat's old installation with new wiring, new ducts, connections or even a new switchboard. By doing it this way you avoid the complexity of designing new circuits, calculating wire gauges and drawing up connection diagrams. The problem is that this can only be done when the old installation is satisfactory in terms of what it can provide.

As soon as you start to increase the demands you place on the system there will be an increase in the technical complexity of the installation. In the case of renovations the time almost always arrives when you become convinced that the best thing to do is just rip it all out, start again and design a completely new installation, which is what we ended up doing.

New installation

In order to illustrate this process more visually we have divided the renovation of the electrical installation into four phases, fully aware that each is intimately connected to the rest, to the extent that, on occasions, they even overlap:

1. **Overall planning**
2. **Layout**
3. **Fitting**
4. **Finishes and testing**

In this chapter we will look at the first two stages. These are possibly the most important in terms of a good final result.

1 - Overall planning

This first phase is mainly theoretical, with more time spent sitting at a desk with pencil and paper than actually doing any work. It is a question of defining and drawing up a list, like a letter to Father Christmas, of the equipment that you want to install in the boat: from the 220 V sockets to the drainage pumps, including multipurpose equipment, light fittings, the battery charger/converter and the windlass.

For *Samba*, on the basis of this list, we consecutively itemised (you don't have to follow any particular order when you do this) each and every one of her new electrical needs, even before we had defined the different makes and models of equipment that we would be installing.

To do this you have to be meticulous, and it is always better to over rather than underestimate, even to the extent of allowing for apparatus that you may not wish to install immediately, but that you might require in the near future. In this way the corresponding wiring and switches can be installed in readiness on the panel. Any large scale errors, or major modifications, are a spanner in the works with regard to the correct development of the succeeding phases.

As the cherry on the cake, on *Samba*, we decided to install a small independent 220 V ring main, powered from the batteries using a solid state converter. This would also be capable of being connected directly to any on-shore 220 V supply and drawing electrical energy from it.

One aspect that you will need to resolve during this phase is how you are going to recharge the batteries. In our case we used the alternator on the engine and/or the battery charger from a 220 V on-shore supply, although there are other alternatives, such as generators, solar panels, wind turbines, the engine shaft or drag systems.

➤ *Planning the new electrical installation is eminently theoretical. You will spend more time behind a desk (or the chart table used as one) than actually doing any work. It is a matter of defining and listing the equipment that you are going to want to install in the boat.*

➤ *Unless the new installation is an exact copy of the original, trying to make it up as you go along and trusting in your intuition, is a guarantee of disaster. Numbered lists of wires, the main wiring diagrams and a layout plan for each element must always be at hand and copies must be kept on board even after the refit is complete.*

2 - Layout

Once you have made up a detailed and numbered list of the equipment that you want to install in the future, you then have to decide where to put it all. This list also leads to questions such as what electrical current will be drawn from the batteries, or what output is required from the battery charger. It seems obvious that the cooling unit should be somewhere near the fridge, for example, or the automatic pilot close to the helm. But should the exterior 220 V socket be in the cockpit or the anchor locker? What would be the best place for the battery disconnectors, the drain pumps, the electronic transducers or even ceiling and wall light fittings?

You have to be absolutely precise when it comes to the position of each element, insofar as this is going to affect the best possible way or ways to lay the electrical cable. When trying to position your electrical sockets and lights it is a great help to draw a plan of the boat, even if this is purely schematic, and mark out the position of each element. You can then use the plan to mark out the position of every light, accessory or piece of equipment, plan out the different electrical circuits and mark out where they will have to run.

As soon as the position of each terminal of the installation has been defined then the margin for manoeuvre in any future changes or extensions will decrease, should this arise you will have to go back over the previous steps and start again.

Once the equipment has been listed and its location decided on, decisions will also have to be made concerning the gauges and lengths of the wiring for each piece of equipment. Without wishing to get too involved in technical matters here, you will have to bear in mind that the electrical wiring must be of a gauge suitable for the safe carrying of the electrical current that will be required by the electrical equipment. This is achieved by consideration of the drop in electrical potential (Volts) along the cable and so is determined by the consumption of the apparatus (current, measured in Amps) and the distance separating it from the batteries: the greater the distance the thicker the wire needs to be.

If these standards are not respected then current bottlenecks will occur. The first consequence of ignoring this advice is that the light or equipment is not going to operate adequately due to the consequent, reduced potential (Volts). But the most significant risk is that the lack of gauge will lead to the wire heating up and possibly causing a fire on board. In the table, at the end of this chapter, the right wire gauges are indicated, depending on both current (Amps) and the distance the wire runs in the case of a 12 Volt supply. The disadvantage of choosing over-sized wires is their extra weight, and of course the fact that they are more expensive.

By calculating the total lengths needed for each of the wire gauges that you select, you can buy the wire in bulk. It is always going to work out cheaper to make a single large order from an electrical wholesaler than buying it metre by metre from the corner DIY. For *Samba*, where the new electrical system is more or less standard, we had to order around 300 metres of wire in different gauges. It is difficult to believe just how much wire you need for a 40-footer from top to toe.

Finally, we must emphasise the primary importance of these first two stages (overall planning and layout). Unlike other areas of the refit, fitting the electrical installation (running wires, making connections, etc.) is generally much easier than the previous theoretical planning stage. With a wiring diagram in the hands of a reasonably savvy amateur, the electrical installation of a boat can be completely renovated. This, for example, is not the case with carpentry, painting or fibreglass, where the practical side is always more complex than the theoretical.

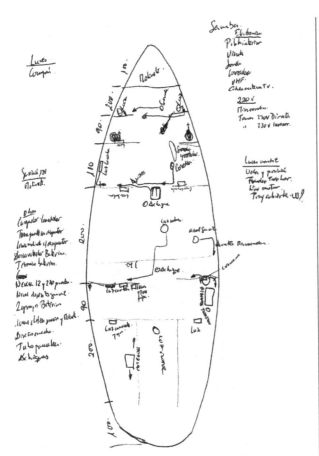

➤ *It is a good idea to plan on paper, even if only on the basis of a rudimentary sketch, where each electrical item or light fitting that you want to install is going to go. When you have this diagram you can then layout the different circuits, precisely where you will run the wiring and the different gauges required for each length of wire.*

➤ *It seemed obvious that the cooling unit ought to go near the fridge and the automatic pilot near the helm, but should the exterior socket go in the cockpit or the anchor locker?*

➤ *You need to be precise when positioning each element, as this will condition the best possible layout for the future wiring. The engine battery (85 Ah) was going to be fitted between the thwarts and close to the starter motor.*

➤ The two 140 Ah service batteries were replaced – once restored – in the space reserved for them on Samba aft of the engine.

➤ We found a space for the battery charger (40 Ah model) and the 12/220 V converter (1,200 W model) in one of the forward cabin lockers.

The wiring diagrams

The most basic 12 volt electrical installation would consist of a wire running from each of the poles of a battery to a light bulb that, in our example, would be permanently on. In order to turn it off you would need to add a switch, and to protect the circuit from overloading or shorting you would have to install a fuse, or a circuit breaker (both elements fulfil the same function). Having added these elements you now have your basic electrical circuit. From this point all we do is complicate things by extending the assembly.

Let's say, for example, that on your boat you need three light sockets, an electric drainage pump, a water pressure unit and three electronic dials. According to the initial plan, a wire would be run from the battery to each of these elements, each with its own on/off switch and fuses. While such an installation would still be basic, you can now get an idea of how the bunch of cables running from the battery round the boat suddenly starts to grow. In order to simplify things here a switchboard can be introduced.

The main function of the switchboard or control panel is to centralise control of the different circuits. Basically there are two wires (positive and negative) running from the battery/batteries to the switchboard. All the different electrical circuits then emerge from the switchboard, each one protected by a main switch and a fuse/circuit breaker, which will cut off the electricity in cases of overloading or malfunction.

In the two cases the concept of the electrical circuit is different. With the interior lights, for example, one or two circuits can be set up with each running round the interior of the boat. In this way, two wires (positive and negative) are run towards the bow and then, by means of an intermediate connection, can branch out to the different lights at the forward end of the boat, each of which would then be fitted with its own switch. The same can also be done at the stern, or anywhere else on the boat that may need a light. It is just a question of creating the different circuits and trying to ensure that each of them has its own operational and installation logic.

The concept of multiple circuits can also be applied to cockpit electronics, where a single wire will carry the current to the different gauges but will not serve for the navigation lights or the drainage pump, which will both need both an independent wire and a switch on the switchboard. These are elements that are switched on individually (that do not form part of a circuit) and it would not be very practical to climb the mast or go down to the bilges every time we wanted to turn the lights or drainage pumps on. These elements must have their own switch on the main switchboard.

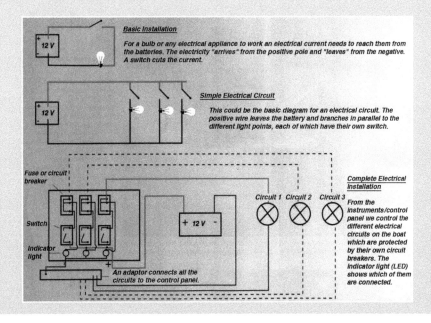

What battery capacity is required

Planning an electrical installation is also the right time to figure out what battery capacity (ampere hours) you are going to need, batteries being the only way to store electricity on board. The choice of batteries will determine how long you can manage to go before you have to recharge them.

In theory it is easy to calculate the current drain (Amps) required from the batteries. All you have to do is sum the electrical consumption of all the installed electrical equipment (see adjacent consumption levels chart), define how long each element is going to be connected during the day and then choose batteries that will give you sufficient capacity to supply your needs for either 24 or 48 hours, the minimum and maximum baselines.

However, in practice this calculation is rather idealistic, and almost impossible to calculate. We can establish an average daily use for the automatic pilot, but it is unlikely that its use will coincide with the use of the mooring light or the anchor winch. For the drainage pump we can anticipate, let's say, five minutes use per day, but if there's a leak this figure would shoot up rapidly, although it would be unusual for somebody to decide to have a shower, use the microwave or the onboard entertainment system while the boat is shipping water and the drainage pump is operating non-stop.

In short, and moving away from pure theory, to resolve this equation for *Samba* we used a calculation (endorsed by experienced professionals) that can be summarised as, the sum of the current required (Amperes) for each piece of electrical equipment on the boat, multiplied by two.

This figure, when rounded up, corresponds to the battery/batteries that will be required to provide that power, for a minimum of 24 hours, without being permanently dependent on starting up the engine or plugging into the 220 V socket when in port.

You must also remember to provide independent batteries for the engine and services, forget about using one to back up the other. The former would be of sufficient dimensions for the engine, which needs to be started while the other/others would have to meet the anticipated requirements for the boat as a whole.

An additional, exclusive, battery for the anchor winch and/or bow thrusters is also an interesting option to bear in mind. These accessories consume quite a lot of electricity and, if they have their battery alongside, this will enormously increase their effectiveness, while also saving on the cost of running thick gauge wire all the way to the bow.

Power Consumption (Watts)	
Bilge pump	60
Pressure unit	60
Electric W.C.	60
Navigation light	25
Mooring light	20
Deck spotlight	40
Electric fridge	70
Interior light	8
VHF radio	5
GPS/Plotter	5
Wind unit	5
LCD screen sonar	15
CRT screen sonar	30
Depth finder	5
Log	5
LCD screen radar	20
CRT screen radar	50
Automatic pilot	5
Compass light	5
Electric winch	1,000
Microwave	800
TV or PC	50
Tablet / smart phone	20
Bow or stern propeller	1,000

This chart is an approximation of consumption levels to establish necessary battery capacity. The figures are for guidance purposes only and will differ depending on the make, model and specific requirements of each piece of equipment.

In order to determine the current drawn by any apparatus, all you have to do is divide its power consumption (W) by the electrical potential (in our case, 12 V). For example, the current drawn by a 12 Volt, 60 Watt bilge pump will be (60/12) 5 Amps.

Having added together the current requirements for all the equipment and multiplied the resulting total by two, you will get a number that is close to the required battery capacity in Ampere hours .

Recommended wire sizes (mm^2) for 12 volt electrical installations, depending on amperage and the distance of the battery/batteries

		Electrical Current (Amperes)									
		2	3	4	5	6	8	10	12	15	20
Distance to batteries (metres)	2	0.5	0.5	0.5	0.75	0.75	1	1.5	1.5	2.5	2.5
	4	0.5	0.75	1	1.5	1.5	2.5	2.5	4	4	6
	6	0.75	1	1.5	2.5	2.5	4	4	4	6	10
	8	1	1.5	2.5	2.5	4	4	6	6	10	10
	10	1.5	2.5	2.5	4	4	6	6	10	10	25
	12	1.5	2.5	4	4	4	6	10	10	10	25
	14	1.5	2.5	4	4	6	6	10	10	25	25
	16	2.5	4	4	6	6	10	10	25	25	25
	18	2.5	4	4	6	6	10	25	25	25	25
	20	2.5	4	6	6	10	10	25	25	25	25
	22	2.5	4	6	6	10	10	25	25	25	25
	24	4	4	6	10	10	16	25	25	25	35
	26	4	6	10	10	10	16	25	25	25	35
	28	4	6	10	10	10	16	25	25	25	35
	30	4	6	10	10	10	16	25	25	25	35

Chapter 6

Repairing the deck (part 1)

As with the interior, to restore the deck you have to start by removing the boat's fittings until she's as bare as the day she left the mould. At this point you can start the restoration work.

Preparing the ground

There were various improvements to be made with regard to the organisation of her deck and fittings, all of which will be broken down and explained in this and successive chapters. But before getting down to business we had to decide on and plan, as far as possible, all the steps that we needed to take to achieve our goal. While obviously impossible to show this initial stage in terms of text and photos it is, nevertheless, an essential one when it comes to ensuring the success of a refit and an aspect that must be studied with great care. Redesigning a seventies deck, updating it to the present century, requires

a great deal of imagination, I can assure you of that.

Fortunately we were able to count on the inestimable help and experience of technical personnel from Lewmar, who made a study of the best or possible locations for the installation of the necessary parts. They calculated the stresses that we could expect from different manoeuvres and the type of cars, pulleys, return pulleys and clamcleats that would be most suitable for *Samba*. The same was done with the new hatches and companionway that we were going to install. You need to know what sizes and models are going to be best suited to your boat, because any modifications that you make to the deck will depend on them.

Although there are a lot of chapters still to go before we will get round to mounting the new deck fittings, most of the parts were already aboard and we would be checking to see how they fitted on the deck as the repair work went ahead.

Trying to buy as much as possible from the same supplier and putting in your order well in advance gives you the advantage of being able to negotiate advantageous prices. But there are two other advantages in planning as thoroughly as possible that also result in cost savings. Having a clear idea of what you are trying

to do makes things run much smoother, whether you are doing the work yourself or getting tradesmen in, and avoids time wasted on experiments or improvisation. At the same time, when your final aim is clear there are fewer opportunities for making mistakes, insofar as you will not have to go over work already done because of changes to the overall refitting plan.

Getting it all one hundred per cent right is impossible but, as the old sailors say: "How can you set the right course when you don't even know where you want to go".

Step by step

➤ Before getting down to the work you have to examine, plan and have a clear idea of the steps to take in order to get the results you want.

➤ We filled a number of crates like this with the old winches, cleats, vent holes, stanchions and dorade boxes. Once we had these on land we made a selection, separating parts that were in good condition (unfortunately not many), cleaning them in case we might need them again and setting them aside. When doing a refit never throw anything away until it has been satisfactorily replaced.

➤ What with the broken parts and those that we were going to have to move, we had to strip down practically all the deck fittings and accessories. In some cases this was relatively easy while in others it was more complicated. For the most resistant parts we had to get out the saw or anglegrinder to remove stubborn screws and bolts. We then covered the holes with tape, as a short-term solution in case of rain.

➤ One of the rigging changes would be to move the mainsheet traveller from its position in the cockpit to the cabin roof. The existence of electrolysis was a complication when it came to extracting the screws attaching this to its solid wood support so, in the end, we had to use a jigsaw to finish the job off on the sail side. For the time being we gained seating space for two on the cockpit benches.

Preparing the deck for new stresses

Before installing any new accessories on the deck (windlasses, cleats, clamcleats, pad eyes, winches, etc.) you have to anticipate the stresses that are going to be generated and apply the necessary reinforcements. If this is not done the resulting problems and breakages could be serious and costly. On the cabin roof of *Samba*, for example, the traveller, halyard clamcleats and new winches represented 200 kg of additional traction that had not been anticipated on the laminate cabin roof of the original North Wind 40, and which required lamination of the interior plywood reinforcement.

➤ *The new halyard clamcleats to be installed on the deck were slotted into a gap made in the drip rail.*

➤ *After taking the necessary measurements, making sure there would be neither too much nor too little space, a section of the drip rail was cut out, using the cutting accessory on a grinder. There was no going back now.*

➤ *In a few minutes the gap had been opened out and the interior reinforcement laminated.*

➤ *From the inside, and using an electrical scraper, we removed the surface layers of fibreglass on the underside along with the remains of old glue and textile linings. If you do not do this the new reinforcing laminate will not adhere correctly in place.*

➤ *This is what the underside looked like after it had been prepared for the new reinforcements. The superficial fibreglass laminates had been removed until only healthy layers remained.*

➤ *The first stage was to laminate four layers of mat and fibreglass which would serve both as a base and cover up the hole made in the cabin roof.*

➤ A handy tip for making sure that the top side of these laminates are flat is to use a piece of wood with a plastic finish, either taped on or held down with a weight. Glass fibre adapts to this flat shape, but will not stick to it and, after it has dried, the wood is very easy to remove.

➤ Underneath the new fibreglass layers we attached 15 mm thick sheets of marine plywood. This added the necessary resistance to the areas subject to new stresses. To make sure that these sheets are fitted and glued perfectly, daub them with fibreglass paste.

➤ The sheet is then fitted into place and held there by planks jammed between floor and ceiling, to keep them in place until the fibreglass paste sets.

➤ As a finish, the interior sheet of plywood will be covered with several layers of fibreglass, as far as the adjacent vertical bulkheads. In this way you achieve greater overall resistance, both horizontally and vertically.

➤ Back on deck, to seal the sides of the holes opened over the cabin, first fill them from the outside using a little fibreglass paste, then laminate some layers of fibreglass over them so that the area becomes more stable and will not split, something that often happens when applying fibreglass paste over extensive surfaces without reinforcing the surrounding area.

➤ Working to set sizes it is always good to make sure the fittings that you install actually fit where they are meant to. In these first stages of the refitting any such errors can be easily rectified. But if you find that a part does not fit once the deck has been painted, then you will have a monumental disaster on your hands.

Chapter 7

Repairing the deck (part 2)

Every refit has its frustrating moments, particularly when the boat that you are trying to improve is gradually being turned into a pile of junk.

Filling holes

At this point *Samba* looked really sorry for herself, both on and below deck. The refit, instead of advancing, gave every impression that the boat was getting worse; it was as if we had entered, of our own free will, a long dark tunnel with no visible exit, and no turning back. All we could do was keep on slogging away to keep our spirits up.

In the previous chapter we commented on the different reinforcements and adaptations to put in place on deck, to support the new rigging to be installed on the cabin roof. We also talked about the need to remove all the fittings that ought to be retired or replaced. When you remove these parts they all leave traces (holes, bases, supports, etc.) which then have to be cleaned up. This work basically consists of filling with paste and/or fibreglass laminations, although each case requires its own slightly different solution, as we will try to explain in this chapter.

Professional experience

Experts in fibreglass laminates, like good professionals of any trade, have amassed a level of experience and know-how that is light years from the level of your average amateur, although this does not mean that an amateur is incapable of carrying out small-scale repairs on the fibreglass of his boat. A good way to start, and this recommendation is equally valid for many other fields of work, is to pay careful attention to what your expert does, accepting the role of an old-time workshop apprentice.

Having seen the job done you can then start to help with some of the basic work yourself, such as filling holes with paste, or perhaps applying small protective laminates, never structural reinforcements, on horizontal surfaces (vertical surfaces are rather more complicated, while working on the underside of any surface is really difficult).

Little by little, as you start to get the hang of it, you can then begin to take on more important jobs, however, the belief that you can soak up a tradesman's vast experience in a short time is a recipe for disaster. A botched job will cost you dear.

➤ *Little by little you have to clean away all traces of the old rigging and prepare the deck to house the new.*

Step by step

Mainsheet traveller

➤ *Relocating the traveller from the cockpit to the cabin roof also means you have to get rid of the traces where it was attached. The first step, using a chisel, is to chisel out the most deteriorated part of the old wooden support, until you have reached completely sound material.*

➤ *In order to ensure that the new laminations will take hold you also have to remove the gelcoat from the surrounding areas. Make sure not to sand down either more or less than required. When doing this kind of work good professionals limit the spillover area as much as possible. Before repainting, all of these laminated areas will require many hours of filling, sanding and polishing. The smaller the surface area that needs to be treated the less time you are going to need to finish the work. All such savings in time will be reflected in the final cost.*

➤ *In the most inaccessible nooks and crannies a chisel is used to remove the gelcoat.*

➤ *Filling the hole left by the traveller starts with a layer of paste reinforced with fibreglass. Once this is dry three layers of lamination are laid over it to provide solidity and stop any crazing (patterned cracking of the coating surface).*

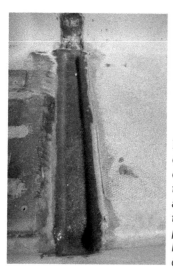

> Given that the depth of the gap to be filled is considerable, following the first layer of paste and subsequent laminations, more fibreglass paste is applied until you have filled it to the level of the bench.

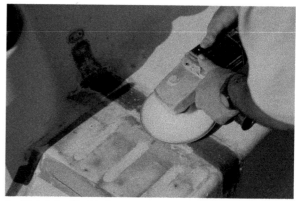

> Finish it off with four surface layers of fibreglass lamination, flush with the horizontal plane of the bench. To avoid lumps, the gelcoat in this area has previously been ground down lightly.

Small holes

> After you have removed all the old fittings your deck is going to end up like a colander, covered in small holes that need to be filled. In small scale applications or filling, laminating will not be necessary, given that fibreglass paste will provide sufficient consistency. In these cases the hole must be cleaned, using a chisel or countersink to bevel the edges, increasing the surface area for adherence and avoiding the paste crazing.

> Apart from that there is no great technical mystery about filling small holes with paste or putty apart from taking care with the putty knife, making sure not to use too much fibreglass paste (more time spent sanding down) or too little (requiring a second application), nor to spread it where it is not needed. All these small details will make a significant difference to the total number of hours of work that you have to put in.

Winch bases

> Many boats have raised and reinforced areas on the deck to support the installation of winches or cleats. In the case of Samba we wanted to get rid of these, mainly for aesthetic reasons.

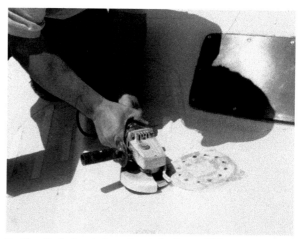

> The first step is to cut out the base. This can be done quite easily using a small electric grinder as a kind of circular saw.

➤ Finally, a few blows with a chisel and the old winch pad lifts off like a lid.

➤ The same chisel can be used to eliminate the top surface of the wood reinforcement that you always find in these areas.

➤ The part that stands out can then be ground down as far as the gelcoat of the surrounding area using a diamond grinding disk so that the new reinforcing laminations can get a grip and will not stand out above the level of the deck. As mentioned above, care must be taken not to sand too much or too little. This will be appreciated when it comes to the final finish and particularly with regard to the time required for filling and sanding before painting.

➤ Before adding the layers of lamination small internal irregularities must be filled out and evened up using fibreglass paste.

➤ Finally four layers of fibreglass are applied, previously cut to shape. At the bottom of the photo you can see one of these winch bases after it has been levelled off and is ready for final filling.

Cleat bases

➤ After cutting off the top cover of the base, as we did with the winch bases, internal irregularities are filled using fibreglass.

➤ Once the outline has been trued, using a diamond grinding disk, which will also remove the gelcoat from the surrounding area, four layers of fibreglass are laminated, cut to size and used to reinforce and level off the deck.

➤ In order to save time, experienced tradesmen alternate the different jobs on the deck and attack on various fronts at the same time. Everything cannot be done straight away, you will have to wait for the resin to set, you will have to measure up the different layers of fibreglass and cut them out, you cannot sand and produce dust that could blow onto laminations that have not yet dried. On the other hand resin or polyester paste has only a limited application time once a batch has been made up, if you prepare more than can be used in that time it will go off before you have finished the job and you will have to throw it away. The experts manage to progress steadily, without delays, by moving from one job to another and avoiding dead or unproductive time.

Delamination problems

➤ Fortunately, among Samba's numerous and varied ailments delamination and rotting wood or laminations were virtually non-existent, which speaks volumes of her original manufacturing quality. Professionals can detect delamination symptoms by the dull sound that the fibreglass makes when tapped with a piece of wood. We found one such damaged area on Samba, about half a square foot, close to one of the old vents, badly sealed, right next to where one of the new halyard clamcleats was to be installed. This had to be 'sorted'.

➤ Loose pieces of wood and fibreglass can be blown away using compressed air. This will also help with the internal drying of the sandwich.

➤ When the whole area is clean, dry and free of loose material, and because the surface area that needs to be repaired is small, it can be filled with fibreglass paste and the piece of fibreglass that was originally cut off refitted. In cases of delamination that cover a larger area, the treatment applied is similar but the filling must be complemented using balsa wood or rigid foam, so as not to overload the deck with the weight of excessive quantities of fibreglass paste.

➤ After cutting out the piece of fibreglass that covered the affected area we also had to eliminate the rotten wood around it. To do so we used punches, various kinds and sizes of putty knives and scrapers. The important thing was to make sure that absolutely no traces of rot or damp remained.

➤ The last step was to sand down the gelcoat surface surrounding this area and apply a couple of layers of fibreglass to reinforce it where the fibre had been exposed.

Larger holes

➤ Filling the old vent holes, which are about 12 cm in diameter, requires a specific treatment. The first step is to apply three layers of fibreglass, working from inside the boat, giving this patch a slight upwardly dished shape. Before doing this you must eliminate the surface fibreglass layers around the hole, in order to improve the adherence of the new lamination.

➤ Up on deck, as with other similar jobs, the screw holes left by this old dorade box are enlarged using a counter-sink, to encourage the adherence of the fibreglass paste.

➤ To improve the adherence of the new laminations and avoid lumps forming on the deck, first eliminate the surface gelcoat using a diamond grinding disk and then fill using a little fibreglass paste to bevel the edges of the hole and ensure contact with subsequent laminations.

➤ Then four layers of fibreglass lamination are applied, in this case dished downwards, until they contact the previous laminations, applied on the interior. In this way, seen from the side, the laminations would form a kind of 'X', providing maximum structural resistance.

➤ The second application of fibreglass paste fills the hollow that remains on deck, leaving enough space for the four final lamination layers, applied to avoid surface crazing.

➤ Finally, on the interior of the boat the underside hollow (previously laminated in a form that was dished slightly upwards) is filled. On the underside, the work does not have to be finished by applying further layers.

Chapter 8

Electrical installation (part 2)

Having defined the electrical needs and the precise location of each item in the boat you are now ready to start installing the wiring. You must always keep lists, plans and diagrams to hand, throughout the refit.

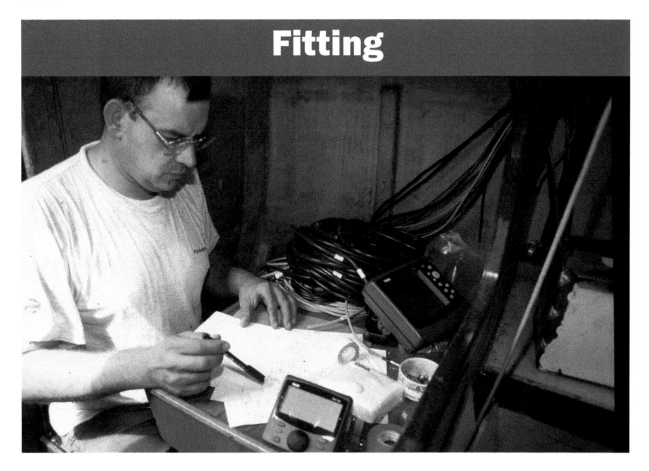

Fitting

The list with the numeration, gauge, length, starting and finishing point of each of the wires serves to definitively identify each wire used on board your boat. Each wire must be marked with its number in felt tip, at both the instrument panel end and the end to be connected to the light or other apparatus. Even after taking all these precautions the possibility of getting mixed up is still very high. Trusting to your memory, however, is a recipe for disaster.

Along with this numerated table, and before you actually start to do the work, you must also have a clear idea of the connection diagrams, as a number of the wires that you are going to install will depend on this.

The numbered lists and electrical installation diagrams, which have been obligatory on new boats for years, are the starting point for our renovation and refit, and will also be the base for any circuit extension work or repairs that may be carried out in the future. These papers will have to be kept on board at all times, with copies also kept safe and sound on dry land.

As well as an overall scheme, the engineers have designed more than half a dozen detailed electrical diagrams of the most complex apparatus (charger, converter, engine alternator, etc.). In total there will be over 300 metres of cable of different gauges: 50 metres of multiwire for the electronics, 30 metres of coaxial cable for aerials and a whole bunch of connections taking electricity from the batteries to the different apparatus.

Paradoxically, and as we mentioned in the first chapter regarding this question, it is technically much simpler to actually fit an electrical installation than it is to plan how one will operate. With a wiring diagram in his hands just about any amateur can set about fitting the installation with a good chance of success.

An interesting way for an amateur, both technically and economically, to set about this is to leave the overall planning to a professional and dedicate his time instead to fitting the wiring in place and connecting it up to the more straightforward equipment (installation companies will usually send their least qualified staff to do this work).

Fully qualified electrical fitters may come back later to supervise the work and possibly to take charge of the more complex connections. Subsequent finishing work (clips, conduits, strips, etc.) is also possible, without a great deal of complexity, for the owner himself.

Making quick progress

In order to gain time and simplify the fitting work, for *Samba* we ran the electronics wiring along with the electrical wiring. Although the repeaters for the electronics would be connected later, a lot of time was saved by using this method to run the wires and fit the transducers, the fluxgate compass sensor, the pilot light wires and cockpit repeaters. Nor should you forget the connections between the switch panel and the foot of the mast (wind unit, position lights, etc.). When the mast is raised these will already be in place and all you will have to do is connect up the cable ends at the bottom of the mast.

One advantage of fitting the electrical installation with the interior stripped out is that the work can be done faster and more comfortably. Anyone who has ever tried to run a new cable through the intricate system of ducts of a production boat knows that this is not always easy and in many cases is virtually impossible.

Instead of running the wires one by one, as we did in *Samba*, shipyards that build production yachts cut and tie off the complete installation at their electrical workshop. Later the head fitter attaches these bundles of wires to the practically empty hull (or the underside of the deck) knowing that when all the furnishings have been fitted into place the required wire will appear wherever it is needed. It would be extremely complicated to use this system when refitting an old boat, insofar as you would need to strip it down to the hull (removing the deck and the bulkheads) and put in an immense amount of preliminary design, to an extent only justified when working on a production line.

If drainage pumps, pressure pumps or lights are already installed (in *Samba* there were hardly any) during the fitting these can be connected up.

This is also the time to start to connect the switchboard, the true philosopher's stone of any installation. Renovation starts with replacing the plywood panel that it is built into, alongside the chart table. The carpenter started the interior restoration work here (an aspect that we will be taking a look at over the next few chapters).

When we ordered the switchboard we provided the manufacturer with our numbered list of equipment, which meant that we could get it supplied all preconnected. This option, offered by some manufacturers, consists of connecting up the switchboard dials and switches at the factory. The switchboard is then delivered to the customer with a regular 'spaghetti' of wires sticking out at the back, the numeration of which coincides with the list supplied. Wire 12 at the switchboard, for example, was our wire 12 on the list and we already knew that it corresponded to the stern drainage pump.

➤ *Between the electrical installation and the electronics we used over 300 metres of different types and gauges of wire. The wires for the batteries or anchor winches come with 50 and 35 mm sections respectively, while the electronic equipment uses fine multiwires.*

Step by step

➤ One of the advantages of running the electrical installation wires with the boat's interior completely stripped out is how fast the work can be done. Anyone who has ever run new wires through the wiring ducts that most production boats are fitted with will know how complicated it can be, sometimes it is almost impossible.

➤ It is very important to number each wire at both ends. In electrical materials shops they sell boxes of colour-coded numbers specifically for this purpose. For thicker wires, and/or those with a protective rubber sheath, it is easier to tape them and write the number on with an indelible felt-tip marker.

➤ We soon had two large bundles running into the (future) switchboard: one from forward (top corner of the bulkhead) and the other from aft (under the chart table).

➤ For the new installation we would use a DIN strip as an intermediate stage. In the top conduit you have the installation wires and in the bottom one those of the panel. It is a question of ordering the positive wires of the installation and the panel numerically, in order to connect them up to each other using the central strip.

➤ One by one the positive wires were connected up to each element. The negative poles will be joined in a pair of large connectors on the forward and aft sides of the panel, from where just two wires emerge that close the circuits. The lists and installation diagrams were on permanent view on the chart table. When the boat is finished and sailing again, these must be kept on board as an irreplaceable technical reference.

➤ When connecting wires you will have to be both orderly and meticulous, although you do not need a high level of technical know-how. Any amateur with good manual skills can take this job on, as long as he has access to the installation connection diagram.

➤ If we compare what the back of the switchboard used to look like with what it looks like now, the improvement stands out a mile. In the original electrical installation the wires were connected seemingly without any order or logic to the back of the panel. Instead of this rat's nest of wires each function has now been numbered at the installation, at the DIN strip and the panel wires. This automatically simplifies any work that needs to be done or any extension to the circuits, while also improving the safety and reliability of the electrical installation as a whole.

➤ From the new switchboard, the true philosopher's stone of the installation, all the connections and services can be checked at a glance.

➤ When running the wires along the bottom of lockers and the back of what would eventually be cupboards they end up forming disorganised bundles. These would later have to be straightened out and tied off for safety reasons, and also to comply with present day technical regulations.

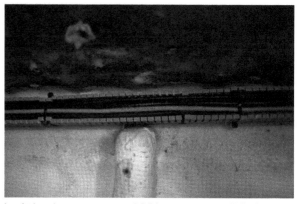

➤ At hardware stores and DIY centres you will find a wide range of ducts, spirals and corrugated tubes, specifically designed to hide and protect electrical wiring installations. Apart from the improvement in the look of the installation it must not be forgotten that every loose wire lying around runs the risk of being accidentally snagged.

A new panel for the chart table

➤ In order to connect up the switchboard we needed a plywood panel to which it could be attached. The system for defining the shape and dimensions of the new panel, very practical for all kinds of irregular adaptations, is making a template, using thin strips of wood, stapled (glued would also do) together.

➤ Then, using this frame as a template, it was not complicated to trace the outline on to a 15 mm thick sheet of plywood. It is also important to have previously marked out on the template a line indicating the precise verticality of the panel. To end up with the wood grain of the veneer sloping at an angle is not very pleasing from an aesthetic point of view.

➤ Before fitting the panel in place, and using polyurethane, the carpenter glued a piece of plywood to the hull. This was to provide a base that the strips and connectors could be screwed on to.

➤ After making a number of minor adjustments the panel could then be fitted into its final location. The carpenter also opened up two holes in the panel; these will serve as bookshelves for the storage of small objects.

➤ The size and shape of the shelves, supported by side strips and the central reinforcing strut, were also defined using strips of veneer to make up a template. This was then traced on to the plywood sheet and the shelf was cut out.

➤ At the back of the bookshelves we glued a wood veneer lining, which we then covered with three good coats of varnish, before finally installing the panel. This would have been much more difficult to do after the panel had been installed.

➤ The first person to work with the switchboard was thus the carpenter and not the electrician, cutting the opening in the plywood panel and mounting the switchboard on to it using a continuous piano hinge, so that it could be easily opened.

Chapter 9
Replacing Perspex side-windows

Replacing Perspex side-window panels with opening portlights is something that many owners would like to do to improve their boats. In this chapter we will explain how we considered this question for Samba.

A breath of fresh air

Many production yachts, in particular those dating from the eighties and earlier, used smoked-finish Perspex panels as cabin side-windows, a simple and economical solution that was adopted at that time by a number of shipyards. Over the years this material gets scratched, darkens, becomes opaque and has to be replaced. So far so good, but with *Samba* we wanted to go even further: we wanted to replace these panels, bolted into place, with opening portlights. This would be a substantial improvement in terms of interior ventilation, one of the main disadvantages of fixed panels.

Catalogue in hand, the first task was to find the portlights that would best adapt to the available spaces on the sides of the cabin, making sure that their location would not interfere with bulkheads, wiring, ducts, vents or any other interior or exterior element. Manufacturers also indicate the size of the openings to be made as required by the frames (information as, if not more, important than the above). It is a good idea to make up a template, a simple bit of cardboard will do, so that you can try the size out both inside the boat and up on deck, checking out all possible portlight positions and even getting an idea of what they would look like aesthetically.

Opening portlights have to be installed on absolutely flat surfaces. If this is not done then the aluminium frame will be forced when the screws are tightened, adapting the frame to the curve of the cabin. This will spring the seals and they will leak relentlessly. Portlight frames should be forced no more than a millimetre side to side, and you will find few boats with surfaces that flat, even those that look as if they are. However, there is no need to panic, if the curve of your cabin is not too exaggerated (let's say no more than 5 mm on each side of each portlight), then this difference can be compensated for using sealant – an easy enough solution and one that will in no way compromise solidity, seal-tightness or aesthetics.

In many cases the curved form of the cabin is almost imperceptible, while it also explains why most boats are fitted with a number of small portlights rather than one big one, which would obviously be cheaper, running the whole length of each side. By having a number of portlights you minimise the arc (the separation at each end) of those installed.

In our case, we found that the best solution was to combine three opening portlights (one of which was smaller than the others) on each side of the boat. However, before installing them, the first thing we had to do was to seal up the old openings and paint them. It will be several chapters before we get to see the final result of this work but the most difficult part will be done in the present chapter.

This is an improvement that can be applied as a one-off job on any yacht. The only disadvantage being that you will have to repaint the deck, or at least the side of the cabin. Similarly, inside the accommodation, you will have to replace, at least partially, the decorative linings.

➤ Installing the new opening portlights starts by blocking the openings behind the Perspex panels.

Step by step

➤ Many production yachts, in particular the ones manufactured in the seventies and eighties, used Perspex panels as windows on the sides of their cabins, this being a simple and economical solution that was adopted at that time by a number of shipyards. Over the years the material gets scratched and darkens, until it becomes almost opaque, at which point it needs to be replaced. However, in our case, we wanted to go a little further than that and replace these panels with opening portlights. In this way we will improve interior ventilation, one of the main inconveniences of fixed panel systems.

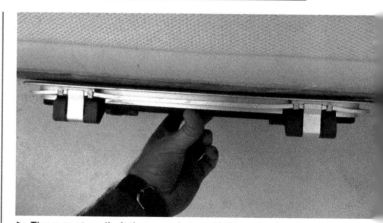

➤ The curvature limit that opening portlights will tolerate is barely over one millimetre from side to side and few boats have surfaces on the sides that are that flat, even when they look as if they are. Tightening up the bolts until the frame is bent to fit tightly really damages the seal. This is the reason most boats are fitted with a number of small portholes, rather than one big one running the length of the side. It is a way of minimising the curvature of their cabins.

➤ First of all we removed the screws and detached the old Perspex panels. Although you can easily break them when doing this (if they are not broken already) it is important to keep the pieces because, as you will see, they will serve as the perfect mould for the subsequent lamination work. Basically all we needed to do was clean off any traces of silicone that was stuck to them.

➤ The refit continued, now working on the inside of the boat, using a chisel or scraper to eliminate the surface coats of fibreglass from all around the frames of the old panels, along with the abundant traces of silicone, glue and the remains of the interior textile lining. If we did not do this the new laminates that we would be applying to fill this gap would not adhere well.

➤ As well as eliminating the surface coats of fibreglass use the grinder, fitted with a grinding disk, to grind down the area around the frames, so that the new layers of fibreglass do not form a ridge.

➤ Work now continued on the exterior, where we eliminated the gelcoat from the outline of the original panels. You can also take the opportunity at this stage to rectify any examples of delamination of the surface fibreglass, caused by the entry of water around the panels. Initially this work is done using a

chisel.
➤ Later the work continued with the grinder, using a diamond grinding disk to smooth down the surface and eliminate the last traces of silicone, the gelcoat and areas of delaminated fibreglass.

➤ The old panels, even if they are broken into pieces, are then reinstalled with self-tapping screws. By doing this, the Perspex will serve as the recessed part of the mould to recover the shape of the cabin when applying the laminate from inside the boat. The wooden batten that you see in this photo is holding in place a broken piece of Perspex that was impossible to screw down.

➤ Back inside, the first coat of resin was applied directly on to the Perspex, to which it would barely adhere after going off. Initially the resin impregnates and maintains the first layer of fibreglass.

➤ These first layers of fibreglass were cut to the precise size of the openings. The application of three fibreglass and mat laminates is equal to the thickness of the frame, which is the thickness of the fibreglass cabin walls.

➤ Once this thickness was obtained the subsequent fibreglass layers were laminated slightly overlapping the original frames. The final result was a consistency even greater than that of the Perspex panels.

➤ When the interior laminate had set, the Perspex panels, provisionally screwed in place on the outside, could be effortlessly removed. The surface that resulted was surprisingly smooth and hardly even needed a coat of putty before it was ready to be painted. However, in order to ensure sufficient overall resistance I recommend that a few layers of fibreglass be applied to the exterior as well.

➤ Before adding the exterior laminates on each side, deburr the fibreglass a little, so that it provides better adherence and minimises any lumps that will later have to be filled and sanded down before painting.

➤ The small holes left on the exterior by the screws, plus any other defects caused by delamination were filled using polyester paste reinforced with powdered fibreglass.

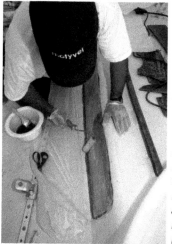

➤ We then laminated three layers of fibreglass, cut to size and slightly overlapping the edges of the opening for the panels and reaching as far as the external reinforced surface. When working with polyester resin, paints or varnishes it is important to protect the surfaces surrounding the work area. This avoids stains that may later prove difficult to remove.

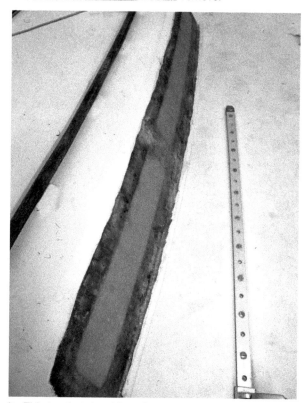

➤ This is what the sides of the cabin looked like after 10 hours' work, with the openings blocked and filled. The most difficult part of the job had now been done but, before fitting the new portholes we had to fill and paint the deck. There are still many chapters to go before we can present Samba's new 'air conditioning'.

Chapter 10

Anchor locker (part 1)

The position of the anchor windlass is something that we agonised over right up to the very last minute. There are different possibilities, each with its advantages and disadvantages.

A support for the windlass

These days nobody would question the convenience of an anchor windlass, especially for anything over a 30-footer or a boat being sailed by people over a certain age. Our decision to install one was taken without a second thought.

The main dilemma was whether to locate it on deck or inside the anchor locker. For *Samba*, and other boats of her generation, where no base was planned for the interior of the anchor locker, installing it on deck would have been far and away the easiest solution.

But the problem was that installing it on deck would have meant that the chain had to pass over the cover of the locker, would have got in the way when it came to opening it and would have made it difficult to get a good view of how the chain was being stowed. At the same time, the almost two metres of chain that would have had to remain on deck would have been both an eyesore and an interference.

As an alternative, installing the windlass inside the locker would ensure a clear and uncluttered area at the bow, simplify the opening of the locker's cover whenever required and would dramatically improve its overall appearance. As you have probably gathered that's the one we went for, in the full knowledge that we would have to make up and laminate a highly resistant support to be fitted inside the locker itself.

Installing a windlass: some technical conditions

In order for a windlass to be effective it must pull on the chain, from the cable lifter to the sheave, in a line as close to horizontal as possible (+/- 5). Such horizontality is not easy to achieve with a windlass installed inside the anchor locker and/or very close to the bow, as is usually the case in small and medium length boats. To achieve this it is necessary to either use sets of sheaves (resulting in a loss of efficiency) or raise the height of the windlass, covering it with a domed (and arguably ugly) locker cover.

On *Samba*, thanks to the size of her anchor locker and the pronounced rake of her bow, we were able to allow for over a metre between the windlass and the bow sheave, while only losing three degrees to horizontal in the run of the chain. In the end we decided on this solution, accepting that the work would be much more complex than if we had decided to simply bolt it to the deck.

Despite the apparent ease with which the laminating specialist did his work, the following photos will give you some idea of what a virtuoso fibreglass construction performance it was. The truth is that, had I been aware of the technical complexity involved (almost four full days' work), I probably would have gone for the easier option,

which was simply to install the windlass on deck. During *Samba*'s refit we made it a rule to follow the dictum, 'simple is best'. In this case, while not wishing to set a precedent, we disobeyed it.

Work on the anchor locker would be completed in later chapters, after the deck was painted, and would include restoring the locker cover, installing the new cleats and, obviously, fitting the new windlass in place. Compared with this first stage the rest was child's play.

➤ *Laminating a support for the windlass inside the anchor locker is a job for specialists in fibreglass.*

Step by step

➤ *Installing the windlass inside the anchor locker is the most complicated of the list of refitting jobs to be done at the bow of the boat. This task included reinforcing the deck to install the new cleats and redesigning the cover of the locker and its support tabs, both of which were suffering seriously from delamination.*

➤ *First we traced out the line of the new locker opening that would have to be cut, making sure that the windlass and the chain would always be on view during anchor lowering and hoisting operations.*

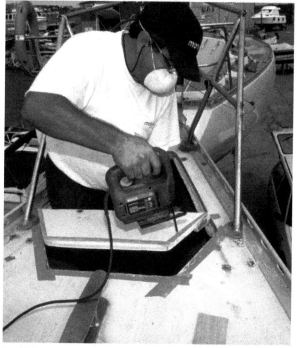

➤ *Using an electric jigsaw we quite simply followed the line and enlarged the anchor locker opening. We left a distance of 40 cm between the line of the cut and the bulkhead that separates the locker from the forward cabin.*

➤ For logistical reasons (it would have been complicated to install these reinforcements after the base had been laminated) the specialist started out by laminating the two plywood sheets to be fitted below deck, which would act as the bases for the new cleats, to make sure that the balsa wood sandwich would not give, under pressure from the screws, and reinforcing the area in anticipation of the new stresses that it would be subject to. The parts slotted into the back part of the locker, even leaving a small overlap towards the bow, which would serve as a flap for the frame of the new locker cover.

➤ The first gluing of the plywood sheets onto the bottom of the locker was done using polyester paste reinforced with glassfibre. After fitting the parts in place, the specialist then clamped them. Once the paste had set they were laminated with various layers of fibreglass which extended as far as the sides of the hull and the bulkhead at the back of the locker. This ensured that the resistance of the new mooring cleats was guaranteed from all angles.

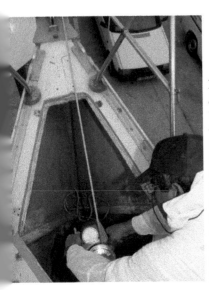

➤ The location of the windlass inside the anchor locker was selected taking careful consideration of the horizontality of the run of the chain in relation to the anchor sheave. The extra-flat Lewmar Concept 1 design almost allowed this to be raised to deck level, while it was also necessary to true up the slight slope of the base inside the locker, so that the pull on the chain was clear of the bow sheave.

➤ Once the position of the windlass had been set the next step, using a diamond studded disk, was to eliminate the paint and the first layers of fibreglass inside the locker around the area in which the new base was going to be laminated.

➤ The specialist then made a template of the shape of the base inside the locker, using strips of plywood stapled together.

➤ Using this template he then traced out the shape of the base on two sheets of 15 mm thick plywood (a total thickness of 3 cm). After making sure that this piece slotted into place inside the anchor locker, the next step was to laminate the support as a block so that water would not penetrate it and break down the resistance of the wood. Rounding off the front edge would facilitate the laminating, while also improving its overall appearance. Once the part had been laminated all that was required were slight adjustments to make sure that the base slots perfectly into place inside the locker.

➤ The new base is then provisionally installed in place, using wooden blocks to wedge it. Now was the time to laminate it firmly to the hull.

➤ Once finished, you can see how close the windlass is to the level of the deck, while hiding the run of the chain beneath the anchor locker cover (which would have to be replaced). The base, laminated to support the windlass, is also slotted on to ribs that already existed on this part of the hull, blocking any possible forward movement towards the bow. It took four days of non-stop work to reach this stage of the installation, giving you some idea of the complexity of a task that at first sight seemed straightforward. Before the windlass could start to operate it would have to be connected to its motor, and this was not going to be done until the deck was painted and the electrical installation completed.

➤ Laminating fibreglass on the inside of the anchor locker was not easy. You often have to work blind and/or in a very uncomfortable position. Just as well Samba's anchor locker was quite roomy! To make the work easier the specialist laminated two layers of fibreglass, sandwiching a layer of mat, on to an external piece of wood. These improvised layers were then moved to their final location inside the locker.

➤ It was laborious and uncomfortable work. In the end ten layers of fibreglass had been laminated on to the sides and the bottom of the new base, both at the top and the bottom. The specialist took care to start off working the underside of the base and, as a result, when laminating the top side, the polyester resin that dripped down between the base and the hull did not run away and was not wasted. It was trapped by the laminating on the underside, filling small gaps and increasing the overall strength of the assembly.

➤ The windlass and its recently laminated base gave a youthful air to Samba's old anchor locker. On the bow side of the locker you can still see the four holes where the old windlass used to be attached. So far forward the windlass barely seemed to have had space to offload the chain, which always got snarled up. In the end the last owner had to remove it all together as it was not working at all.

Restoring the anchor

While the anchor locker was being updated, we decided that the best thing to do would be to restore Samba's anchor, a 35-pound CQR, famous for getting a good grip on all types of bottoms. The conservation of this anchor appeared simple enough but, what with the accumulated rust and flaking coats of old paint, our anchor was looking very much the worse for wear and in good need of some 'beauty treatment'. Nobody likes to have an anchor, which is always on deck and in full view, that lets down the rest of the boat. Luckily, however, our anchor had not been badly knocked or bent out of shape, which would have made restoration much more difficult. In these cases most experts tend to recommend immediate replacement.

Iron and steel are relatively porous metals on which rust and paint become firmly encrusted and are then difficult to get rid of completely. If the surface to be cleaned is small and flat you may have some success using a wire brush, but on an anchor, with its etched lettering and nooks and crannies that are so difficult to get into, the simplest solution is just to take it to be galvanised. This is the recommended solution when your anchor is in an advanced state of degradation.

Whatever the case, taking a small 35-pound anchor to be galvanised in a zinc tub measuring 10 x 2 x 2 metres is an industrial aberration. To get a good price the best idea is to take the complete anchoring assembly (anchor and chain) and not to pressure them too much in terms of deadlines, in the hope that your anchor and chain can be included along with a much larger order. Some slipways offer this service to their customers and it will save you having to lug anchor and chain off to some distant galvanising workshop.

The process of protecting iron or steel parts by immersion in a tub of molten zinc is called hot-dip galvanising. Zinc oxidises very slowly and, without changing its silvery colour, which gradually just loses its original shine. The galvanising process includes previously cleaning off grease, old paint and rust, by immersion in baths containing acids and degreasing agents, thoroughly cleaning out the pores in the metal, so that the zinc will take firmly and in depth inside them, to form a protective coat.

Hot-dip galvanising is a relatively cheap treatment often used for parts that have been exposed to the elements (grills, lamp posts, roadside railings, etc.). If the work is well done it will be effective for over 20 years, although this may be halved in a marine environment. Galvanising will also serve as a good base for an additional coat of paint.

Getting back to the anchor, galvanisation will, little by little, lose its thickness as a result of oxidation, but also and mainly as a result of the knocks and scrapes that are so typical in the life of an anchor. In time (6 to 8 years maybe), the reddish oxide of iron and steel will start to come through once more. In the early stages this can be hidden with a coat of paint or zinc sprays, but this will only be a provisional solution and, in the mid term, the best thing will be to take your anchor back to be galvanised again.

➤ *Samba's anchor was really a sorry sight. To eliminate rust and old paint you can fall back on all types of wire brushes and chemical products, but in the end nothing is better than a good galvanisation.*

➤ *Hot-dip galvanising consists of immersing iron or steel parts in a large tub of molten zinc, which penetrates into the metal's pores and forms a coat that protects against oxidation.*

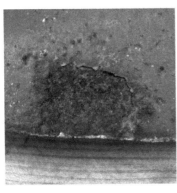

➤ *The marine environment is very aggressive for ferrous metals, which must be conscientiously protected.*

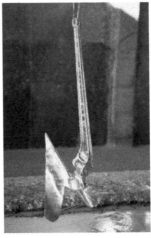

➤ *The change in appearance of the anchor after galvanisation was radical. It looked new! Zinc is a metal that oxidises without changing its silvery colour, just gradually losing its original shine.*

Chapter 11

Repairing the deck (part 3)

Before you can start on the painting you have to restore the hull and deck imperfections, preparing the ground for fitting the new accessories. Once the painters have started, there is no room for these kinds of jobs.

Hatches and through-hulls

The Perspex of the old hatches on the cabin roof was badly weathered and, with time, had turned completely opaque, as is the case in all boats of *Samba*'s age. Replacing these panels is not particularly complicated, you just have to take out the sheets of Perspex and replace them with new ones, cut to size. Neither is it exactly a Herculean task to renovate the seals, replacing the old seals, which have dried out in the sun, with new watertight ones. Taking the frames away to be polished and/or re-anodised is not particu-

larly expensive either. But then there are the fasteners and hinges, which need replacing (most of these were broken), and these parts now have to be made to measure; the company that made *Samba*'s hatches went out of business some years ago, and it all starts to add up.

The 'fix or replace' dilemma is one that will arise frequently during any refit. In our case, bearing in mind that the investment required to fix the old hatches was not far short of the cost of replacing them with new ones, this was an option that would also provide a modern technical solution and look aesthetically pleasing. Moreover, replacement, in the case of two of our three hatches, simply consisted of removing the old ones and putting in new ones. In short, replacement was far too tempting a solution to be ignored.

In this chapter we will also be illustrating our solution for adapting the other new hatch (the one that did not match the dimensions of its predecessor) to the deck and building up a flat surface on to which it could be fixed once the deck is painted.

Another aspect that we will be finishing off in this chapter is the stopping and laminating of the holes left by the through-hulls that no longer serve any purpose or have changed place, a very common situation during any refit. With this work we shall be leaving deck and hull practically ready for painting.

Step by step

➤ The hatches have to be installed on flat surfaces pre-formed for this purpose on the deck. If not then the slight curve of the cabin roof will compromise their watertight sealability. The problems arrive when you cannot replace the old hatch with one the same size. In this case it is always a better idea to install a bigger hatch rather than a smaller one. Using a smaller hatch would be much more complicated technically and would tend to be less aesthetically pleasing.

➤ There are no great secrets about removing an old, broken hatch, the stainless steel screws are all visible and, in our case, none of them turned out to be either broken or seized. Although we must make one important proviso: so far, in this book, we have avoided discussing the kind of problems that you are likely to come up against when stripping down old parts. Inevitably you will spend many a frustrating hour trying to remove rusted or inaccessible screws, screws with broken heads and screws that break when you try to force them. How much time you will waste in this way is an imponderable, but you will have to allow for it.

➤ The first step is to trace the hatch frame onto a sheet of plywood with a Formica-type finish, leaving interior and exterior clearances of some 5 mm so that the rounded shape of the corners on the base does not later affect the bottom frame. You could always do this work in advance, taking your measurements from the manufacturer's catalogue or website, although it is always prudent to wait until you actually have the new hatch before you start.

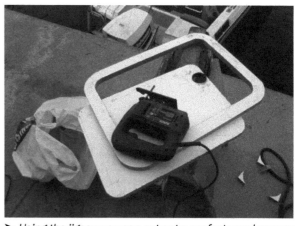

➤ Using the jigsaw you can cut out a perfect wooden copy of the hatch frame. You could, of course, miss out this stage and use the hatch itself to mark out the shape of its frame on the deck, instead of this plywood template. However, apart from the difficulty of handling the hatch as a template, by following this shortcut you are bound to mess the frame up with scraps of fibreglass paste.

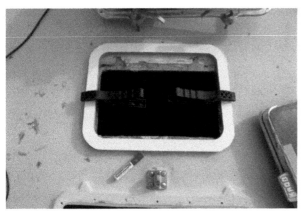

➤ After clamping the plywood template firmly to the deck, using clips or clamps, the next step is to trace the line of the interior cut on to the deck, subsequently cutting it out using the fretsaw.

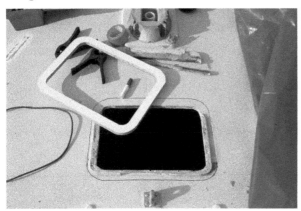

➤ With the interior cut defined, and also using the template, trace the external silhouette of the frame, the one that marks out the exterior outline of the base.

➤ In order to ensure that the fibreglass paste will adhere as best it can to the deck, using a diamond polishing disk, eliminate the gelcoat from the whole surface occupied by the new base. On the sides of the opening we have kept the remains of the base from the old hatch, which will serve as a horizontal reference. Without this reference we would have inserted four self-tapping screws in line into the deck, to act as a support for the template.

➤ Before starting work we taped round the outline of the base to avoid getting it messed up with fibreglass paste, which is awkward to clean off later. Although there is scarcely any adherence between the Formica and the polyester paste we enclosed the template to reduce this sticking even further. The same effect can be achieved by spraying lubricant on to the wood or simply by applying Scotch tape to the bottom side of the template.

➤ After applying abundant fibreglass paste around the edge of the opening it is time to put the template back in place. The height and horizontal level of the base are retaken using the remains of the fibreglass that we kept from the old opening, although the template now marks out the outline of the new hatch.

➤ Clamping the template firmly into place, carefully fill the exterior gaps and eliminate the excess fibreglass paste from the surrounds. The better the initial finish the less time we will need to spend filling and sanding it down. This is essential but laborious work that must be completed before the deck can be painted.

➤ A little fibreglass paste applied to the inside edge will protect the opening from the occasional ingress of water, which could rot the balsa wood used to fill the compound sandwich of the deck.

➤ It is better to remove the tape used to protect the area around the template before the fibreglass actually sets. If this is not done, a firm bond could form between the tape and the fibreglass, complicating its subsequent removal.

➤ On the following day you can remove the template: this is easily done, you just have to insert the blade of a scraper and prise it loose.

➤ About three hours' work should be enough to finish preparing the base of the new hatch. All you have to do now is round off the edges and fill in those small imperfections which are always left by fibreglass paste. Once it has been painted, the base will look like it has always been there.

Filling holes in the hull

➤ Samba's hull had been perforated by the holes made for the through-hulls used in the different layouts that had been installed on board during her long history. Following changes in the location of the head or the drainage pump outlets, nobody had ever bothered to fill in the old holes, just leaving the old valves in place, many of them in a worrying state, to plug the holes. Some of these old holes will remain and other new holes will also be made. For the moment our concern is to fill those holes that are no longer of any use.

➤ This, for example, is the original through-hull that provided water to the WC, and was installed at Samba's bow 30 years ago (the WC was later moved amidships). The danger of entrusting the boat's water-tightness to this mass of rusted metal, topped off with a cork bung for good measure, hardly needs starting.

➤ Having removed the through-hulls and seacocks the laminator starts the work of sealing the holes by recessing the gaps using a chisel: this will create enough space for the subsequent laminates.

➤ After opening up the holes the next step is to clean up the fibreglass using a diamond grinding disk, smoothing off the areas where the new fibreglass laminates are going to be applied.

➤ It is also important to clean out the through-hull holes using a mandrel or file, to eliminate any traces of silicone that could prevent the firm adhesion of the fibreglass.

➤ Using this straight-edged piece of wood you can get an idea of the gap that needs to be filled by the successive laminates, until the boat's smooth lines are recovered. If the area around the hole were not recessed the repair would create a protuberance.

➤ The fibreglass and mat laminates are earlier cut to fit the area to be patched.

➤ Starting with the smallest sections (virtually the size of the hole itself) and applying increasingly bigger laminates, the hole starts to be filled.

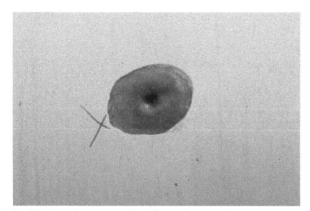

➤ The first layers, applied from the outside, are given a concave form. This means that they will make contact, in the form of an X, with the laminates applied in a concave form from the inside, ensuring maximum overall strength, given that the fibreglass is joined in the area of maximum stress, minimising the subsequent risk of delamination.

➤ Once the first interior and exterior layers have been applied and have set, in contact with each other and forming an X, the gaps that remain inside and outside the hull are levelled out using polyester paste reinforced with fibreglass. When the paste has gone off, five layers of glass fibre are laid over it to provide overall reinforcement.

➤ In order not to waste time while waiting for the laminates to go off, fibreglass professionals alternate the work on the inside and outside of the boat, and also by filling more than one hole at a time.

➤ When you have finished the same straight-edged piece of wood will serve to verify that the hull has recovered its smooth lines. The small imperfections that remain can be later smoothed over with filler to provide a perfect overall finish. Although these fillers, whether polyester or epoxy, can be used to cover over small imperfections they will provide little structural strength, an aspect that can only be resolved using reinforced laminates. On the bottom of the hull this consideration is of crucial importance.

A thirty-year-old mast does not necessarily have to be retired. By maintaining the same extrusion and replacing all of the broken or obsolete components it may still be possible to give it a second life.

Turning the old into new

The first question is obvious. Is it worth restoring a 30-year-old extrusion or would it be more reasonable to think in terms of a new mast? If the aluminium of the mast has not cracked, if there are no dents, no kinks, and if it has managed to avoid the worst effects of electrolysis then, generally speaking, you can say that it is a mast worth keeping. If you are going to update the rigging, then replacing broken, obsolete or inadaptable fittings can ensure that your old mast will continue to function for many years to come.

Another thing would be to go beyond a mere renovation and try to improve the boat's performance by fitting a lighter (perhaps carbon) and/or slimmer mast (possibly with more spreaders). For the *Samba* refit we came down on the side of economic pragmatism, dismissing the idea of investing in a new mast when it was possible to restore and refit the old one at a tenth of the price. Furthermore, the quality of *Samba*'s extrusion, the connectors and masthead, all of which were standard in production yachts 30 years ago, are now the reserve of top range manufacturers. And, if worst comes to worst, this is something that can always be done later on when the flood of refit costs has started to abate.

With regard to her boom, the professionals that we talked to were of the opinion that it was not worth even trying to bring it up to date. The extrusions from the sixties and seventies (as is our case) barely allow for any improvement at all. They will not allow reefs to be taken in from the cockpit, nor hauling on a cunningham or downhaul or even the installation of a rigid vang. Our refit was going to require the installation of a new boom.

With reference to *Samba*'s standing rigging, which was actually replaced about eight years previously and, on the basis that running an x-ray check, the only truly reliable method, would be more expensive than replacing it, we decided to keep it. The diameter was more than sufficient and the components were good quality. The experts were of the opinion that it should still be good for a few years yet.

By the time we had done the work described in this chapter the mast was almost ready to be raised, yet there were some jobs that we decided to postpone, such as installing the new boom, replacing the halyards and cleaning the mast. These delays were merely a question of common sense, given that it would be many months before we could reinstall the mast, which would only get dirty again, covered in the dust of the slipway. The renovation of *Samba*'s mast would culminate with the replacement of the electrical wiring and electronics, other questions that we will be getting back to in later chapters.

➤ The original rigging was a reflection of the specifications and technical possibilities of 30 years ago, when modern pulleys, deck-mounted halyard clamcleats, genoa furlers or high module halyards hardly existed, equipment that has revolutionised the rigging design of all kinds of yachts. Having modernised the deck we now had to adapt the mast to the new requirements. Fortunately, the extrusion was still in good condition and we could still use it.

➤ The boom, on the other hand, could not be updated; its design had been rendered obsolete and would not allow reefs to be taken in from the cockpit, hauling on the cunningham or downhaul, or even the installation of a rigid vang.

Step by step

➤ As with the deck, our first job with the mast was to remove all the fittings that no longer served any useful purpose (such as the mast winch in the photo) or were broken. When removing old rivets all you have to do is drill the head out and then push it inwards with a punch. Nearly all rivets can be removed this way without too much difficulty.

➤ But removing accessories that have been in place for a number of years can still bring unpleasant surprises. What with oxidation, electrolysis and salt residue, a number of bolts tend, literally, to fossilise. If the part or accessory to be removed is not going to be reused (as was our case) you can always cut them off. The problem is, however, more serious when you are hoping to reuse the part. In these cases the experience and manual skills of an experienced tradesman are worth their weight in gold.

➤ The eyestrap for the vang, tightly bolted to the mast, is of no use if you are going to install a rigid vang.

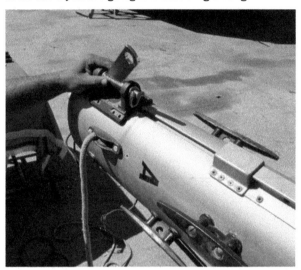

➤ In its place we will install this articulated mast bracket, firmly attached with stainless steel rivets.

➤ The old boom mounting plate on the mast, with a rail that worked as a kind of cunningham, also had to be replaced as it could not be adapted to the new boom that we were going to install. This system was used when the only available sails and halyards were made of polyester and did not allow the mainsail to be firmly drawn. With high modulus materials it is easier to install a cunningham, with a set of pulleys at the base of the mast.

➤ This is what the new mast boom bracket looks like. In the foreground we can see some of the traces left by the work on updating the mast. We have used new rivets to fill the holes left by the old rivets. This may not be the most elegant of solutions but, once the mast has been cleaned, at least the holes will be quite well disguised.

➤ Installing new halyard exit plates is one of the most important aspects in terms of updating the mast. The old system, a very complicated design, formed a kind of chicane of two pulleys, which guided the polyester/rope halyard. What with the salt residue, oxidation and wear produced by the rope, none of these sheave boxes were actually working as they should have been. Fortunately the solution is really simple ('simple is best' as the English say). Instead of trying to fix them we just replaced them with stainless steel exit plates, like the one you can see on the right in the photograph, covering the holes left by the old sheave boxes and allowing the halyard to emerge directly with hardly any friction.

➤ When replacing the halyard sheave boxes, start by removing the old fittings. This is easily done just by drilling out the rivets.

➤ Using the new exit plate as a template first mark the holes for the new rivets with a pencil, and then, using a punch, make a mark on the mast so that the drill bit will not skate off when you start to drill the hole.

➤ The next step, ignored by many amateurs (and by a number of professionals too) is to coat the rivets with a jointing compound. This will prevent electrolysis, created by contact between aluminium and stainless steel, from raising its ugly head over the passage of time, attacking and weakening the weaker of the two metals, which in this case is the aluminium of the mast.

➤ After fitting the rivets, the new halyard exit plate is ready to be used.

The tips of the spreaders

➤ Some amateurs have got into the habit of taping the tips of the spreaders, the part where the shrouds rest, using duct tape. The theory is that this is the best way to protect them from environmental wear and tear. It is a bad theory, to the point of being counterproductive. Instead of protecting the spreader, the duct tape actually prevents correct ventilation and encourages the accumulation of corrosive salt residue underneath it. The accompanying photo is an excellent illustration of the consequences.

➤ Salt had accumulated in an alarming way on the tips of the spreaders, up to now wrapped in duct tape. Fortunately the corrosive effect had not yet had time to attack the aluminium to the degree where its integrity might be compromised. In order to protect the sails from rubbing against the spreaders and the tips holding the shrouds, the best solution is plastic, leather or metal sheathes. Whatever method you choose it is important to ensure that there is sufficient ventilation, that the protected part can be kept clean and that it is easy to inspect.

Setting up the masthead

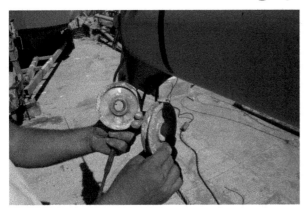

➤ The first job to be done at the masthead is to take out the halyard sheaves. In comparison with the fragile plastic sheaves fitted to most production yachts nowadays it was an absolute luxury to find that Samba's sheaves were of solid aluminium. Time, however, had left its mark, in the form of electrolytic oxidation and some of the interior bearings had also broken. Fortunately this could all be fixed.

➤ After polishing and anodising them again we also renewed the plastic bearings on the shaft. These bearings have a double function: on the one hand they help the sheave to turn and, on the other, they prevent the aluminium in the sheaves from coming into contact with the stainless steel bolts. Once in place the 'new' sheaves can continue to provide good service for years to come.

➤ Another job that needs doing at the masthead is to modify the support for the spinnaker halyard pulleys. The old design, with brackets that emerge laterally from the mast, prevented free movement of the shackles holding the spinnaker pulleys and, in the long term, would compromise the integrity of the base plate itself.

➤ The solution was to cut these brackets off and to weld new brackets on to the top. In this way the pulleys can turn freely, even beyond the diagonal, an issue of particular importance for boats such as ours, where we were going to be using an asymmetric spinnaker.

Chapter 13

Cleaning the winches

Stripping winches down and greasing them is a maintenance task that can be handled by just about any amateur. While winches play a very important role on board they require very little care to keep them in good operational condition for years on end.

Pulling power

In our case, bearing in mind that the deck was soon going to be painted, we set about cleaning and greasing the winches by removing them from their bases. This would help to make the painting and refitting process easier. Removing the winches also meant that we could get in and clean up the wooden bases that they were bolted on to.

This procedure is relatively unusual, insofar as people normally strip down the winch (drum, bearings, shaft, gear wheels, etc.) while leaving the casing bolted to the deck. But whichever way you do it there

are two pieces of advice that professionals never tire of repeating to any amateur who wants to have a go at this.

The first is to remove and/or strip down the winches one by one, putting the different parts in an orderly manner somewhere clean (a rigid plastic box is the perfect place). If you start to strip down two or more winches, even if they are exactly the same make and model, you run the risk of mixing up their parts. Over the years all gear wheels, shafts, bearings or washers become slightly misshapen, due to the particular stresses and wear and tear that they are subjected to. As a result it is recommended that they should always be put back together with their original parts. Stripping down and reassembling the winches one by one also avoids the classic problem (for amateurs) of finding bits that are 'left over' after the work is finished. This being the case, and as winches usually come in pairs, you will always have another one to hand that you can use as a reference.

The second piece of advice is, whenever you can, to clean and grease the winches on land, somewhere clean and well lit. You can do this on board but then you run the risk of making a mess of the deck (grease, diesel, lubricating oil, etc.) or losing one of the parts over the side. By taking the winches on to dry land the only parts that need to be cleaned on board are the bases, which are firmly bolted to the deck. When cleaning the base, by the way, you must avoid the temptation of using wire brushes or steel wool to eliminate any accumulations of salt residue. The metal of the winch may come up as bright as a new pin but the very next day thousands of small rust stains will appear like magic all over the deck; a result of the filings that you have brushed off.

Although all winches are very similar in technical terms each make and model has its own specific parts. Getting hold of the winch manual will be an indispensable guide for stripping it down, and will also illustrate the correct way to put it together again, particularly if you have never done this kind of thing before. It will also serve as a reference, in case you need one, for ordering replacement parts. New boats are delivered with manuals that explain how to strip down their winches. If you do not have one, then the Internet is often an effective way of finding and downloading old winch catalogues.

Step by step

➤ Taking advantage of the fact that the deck is soon going to be painted we started the work of cleaning and greasing the winches by removing them from the deck. This also gave us the opportunity to clean up the wooden winch bases that support them. Sophisticated tools are not usually needed to strip down a winch, although in our case we had to make up an improvised pin spanner (a piece of handrail with two screws) as the original wrench had been mislaid somewhere aboard and was not found until it was no longer needed. Typical!

➤ On land we stripped down the different parts, placing them in the trough where they are going to be cleaned. You should always remove, strip down, clean and grease your winches one by one. By mixing up the parts of two or more winches, even if they are exactly the same make and model, you run the risk of using the parts of one winch to reassemble another. Over time the gear wheels, bearings or washers become slightly misshapen, due to the particular stresses and wear and tear to which they are subject. We recommend that each winch be put back together using only the parts that belong to it. Cleaning the winches one by one also avoids the classic problem that when they have been reassembled you suddenly find leftover parts. In this case, and as winches usually come in pairs, you will always have another winch, as a reference, in case of doubt.

➤ After the winch has been separated it is time to take it on to dry land so that it can be cleaned and greased. This work could be done on board but you would have to take a great deal of care not to make a mess (grease, diesel, oil) and stain the deck. When working on the boat you also run the risk that parts could fall overboard. The best thing is just to take the winches onto dry land and work on them somewhere clean and well lit. This means that the only part of the winch that has to be cleaned aboard the boat will be the base.

➤ Few amateurs will have a professional trough, such as the one in the photo. A good idea for amateurs is to fill a bucket with non-flammable parts washer fluid and leave the whole winch submerged in it, in one piece, to soak overnight. This will dissolve the toughest, most hardened traces of grease and the final cleaning, using a stiff-bristle brush, will be made a whole lot easier.

➤ As you clean the winch you should start stripping it down, piece by piece. Placing the grease-free parts to one side, in a plastic box or on a clean work surface, so that they can dry out.

➤ To avoid problems when you start to put the winches back together again, it is important not to be in a rush. You have to respect the precise location of each washer and gear wheel, even with parts that look exactly the same. Rather than cleaning the whole winch in one go, a good system is first to strip it down and then to clean and grease the parts in independent groups, which can then be left pre-assembled and ready for final assembly.

➤ Running into a reluctant screw or bolt is a common problem when working with equipment that has gone several years without being stripped down. On this winch we ran up against a bolt on the top plate that just would not shift. Rather than risk deforming the slot, breaking the head or damaging the plate itself (it is not always easy to find spare parts for old winches), it is far better to use an impact screwdriver. This tool transforms the force of a hammer blow into a slight but firm turning motion. A force that few reluctant screws can resist.

➤ The ratchet pawls are among the most delicate parts of any winch, both in terms of how it operates and maintenance. When stripping ratchet pawls down you have to take great care to make sure that the small springs that articulate the pawls do not fly off and get lost. The same kind of care has to be taken when putting them back together again.

➤ When this part has been cleaned and reassembled just a touch of oil is enough to keep each pawl operating without seizing. It would be a big mistake to use excess grease or oil when working on winches, in the belief that this is the best way of protecting them. Do not use any old oil for this work; there are greases and oils that will not adhere to the gear wheels, that will not resist heavy workloads, that will dissolve in water, that will emulsify with salt or that will do all of these things at the same time. The best thing to do is to only use the products specified by winch manufacturers.

➤ *For the gears, a little grease on their teeth is enough to ensure good lubrication. Before applying the grease with a brush you have to clean and dry off of the diesel, or other solvent, that you used for the degreasing. If you do not do this the lubricant will become contaminated and lose its adherence.*

➤ *Finally, and using a fine brush, apply just a little grease to the interior drum gear.*

➤ *The main drum bearings, after they have been cleaned and dried, must also have a little oil applied to them before they are reassembled.*

➤ *Once you have cleaned and greased the drum and it is back in place, the winch is once more ready to go to work. This particular job took us two hours, and half of that time was spent freeing a particularly difficult screw. If maintenance had been more regular (ideally this would be done once a year) we would not have had this problem.*

Painting the bulkheads and bilges was our first positive step in Samba's interior restoration. Up to now all we had done was destructive, stripping out the old decoration until we had left her completely bare.

A touch of colour

Sanding down, filling, priming, painting, filling again, sanding and painting... Painting the interior of a boat is hard work and sometimes seems endless. Even when wearing the best breathing and eye protection the dust still gets everywhere, there is no escaping it. The same could be said for the smell of fresh paint which, confined inside a small boat, can make you just as sick as any storm at sea.

But let's start at the beginning. As we saw in the first chapter, Samba's interiors were an amalgam of all kinds of finishes, from damp-swollen wood veneer

to traces of paint in a wide variety of colours, with several areas badly affected by damp, oxidation and grease. So the first thing we did was get rid of all of that, leaving the boat clean, empty and ready to be restored.

The next step was to decide what kind of finish we wanted, both from an aesthetic and a functional point of view. We had a number of options to choose from, the first of which was to redo the wood veneer. This would have been relatively cheap and simple to do, with the one disadvantage that it ages so badly. Damp would surely have raised its ugly head again in the short term, blackening and deforming the new panels.

Another solution was to cover the bulkheads with a fine (4 mm) layer of marine plywood. This solution, which was right in there among the runners up until the very last moment, provides a resistant finish, looks good and can later be varnished or painted. The main disadvantage was the tails or joints between the new material and the boat's existing mouldings, which would have meant practically starting from scratch.

We also considered the possibility of covering the bulkheads with some kind of textile finish (padded vinyl) or Formica, solutions that we eventually used for the ceiling and the shower cabin respectively, as you will see in subsequent chapters.

American style interior

In the end the chosen solution for providing a new shine to *Samba*'s interior was to paint the main bulkheads white. There is nothing new about this solution, far from it. Painted finishes, particularly using white, were par for the course for most boats at the beginning of the last century and lasted practically until the introduction of fibreglass on the scene in the seventies. From then on, and also with thanks to the increasing quality of plywood and varnishes, most production boats began to appear on the market with their interiors lined with varnished wood.

This style of finish, featuring a contrast between white paintwork and varnished wood is known in Spain as a 'Traditional' or 'A la americana' style and still has its fans. It is clean, resistant to a marine environment and relatively easy to renew. White also improves the sense of spaciousness, a question of particular importance in boats such as our dear North Wind 40, which are not exactly renowned for their interior volume. Following decades in which varnish finishes have taken over, today you can even talk about a revival of white paintwork for the decoration of production boats. Fashion does, after all, tend to move in cycles.

At the same time, we also took advantage of the visit by the painters to clean up the bilges, lockers and the interiors of what, in the future, would be her cupboards. After a thorough overall cleaning and degreasing we ended up using as much as 12 kilos of paint on this job. While carpet and wood finishes for the interiors of the cupboards can be a very attractive solution, particularly when you see them at boat shows, the alternative of painting them may seem less luxurious, but it is a system that is cheap and easy to apply, does not soak up bad smells and is very simple to maintain.

Everybody knows that any paint or varnish will only adhere well to surfaces that are clean, dry and have already been degreased. But eliminating all of the old paint, accumulated damp, rust and rot on a boat with a quarter of a century of active service behind it would be no more than a pipe dream. This is why the priority for the paint that we used for the bilges was its ability to cling to the surface and cover it, at the expense of tone or the smoothness of finish.

Working with interiors that have been completely stripped out simplified the work for the painters, even knowing that at some point they would have to retouch the occasional scratch or scrape inevitably left by the carpenters.

The work that we will be looking at in this chapter required about a hundred man-hours; of these perhaps five per cent was actually spent applying the finishing coats of paint. The rest was a matter of cleaning, filling, and sanding all of the surfaces various times until a good, adherent base had been achieved.

This breakdown of time is commonplace for any painting or varnishing job. The materials and the final coats barely add up to ten or fifteen per cent of a professional painter's bill. Amateurs can also help out and try to reduce costs in the early stages, where the work is hardest in physical terms but the least demanding technically. This will concentrate your investment in quality paints and a professional finish, a good way to try to keep the budget down without compromising the final result.

➤ *Before starting to paint you have to eliminate all traces of glue, grease and dirt, as well as preparing that part of the old paint in good condition. To start with, use paint stripper and a hand-scraper, to take off the worst of it, then different stages of sanding, first using a roto-orbital sander, then just the orbital and finally, when finishing off details and nooks and crannies that are hard to get at, using a fine-grain sanding sponge.*

Step by step

➤ Behind a coat of brown paint the seat backs also hid more of that ubiquitous veneer, which we eliminated using an electric paint-scraper.

➤ It was the same again with the undersides of the seats, where we preserved the veneer at the corners, which were fortunately in good condition. If not we would have to have had these parts remade.

➤ Once the old paintwork had been prepared and smoothed down, with all traces of glue eliminated using paint stripper, the backs and undersides of the seats were ready to be repainted.

➤ After smoothing out the most obvious imperfections with filler we used a roller to apply a coat of sealing primer to all of the surfaces being restored. This serves a double function, on the one hand it provides a firm undercoat on top of which to lay the following coats while, on the other, it enhances the imperfections, making them easy to see, which can then be filled and sanded again until they disappear.

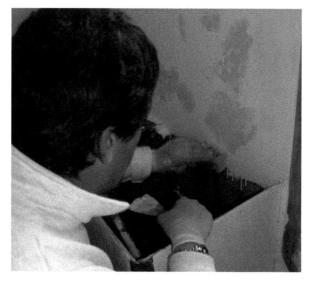

➤ When the primer has dried we have to go back once more and fill holes, scratches and imperfections. The polyester filler is applied using a scraper and has a texture like toothpaste. It sets quickly, in about five minutes depending on ambient temperature, which means you have to prepare small batches so that it does not dry before you have had a chance to use it. It is important to clean and dry the scrapers after each use otherwise the hardened traces of filler will ruin them.

In some areas we also used epoxy filler, which has a slower setting time than the polyester and can be worked for longer. You use the epoxy filler when you need to apply it more thickly or when the filler has to be more resistant. It has a very smooth finish but is harder to sand down than polyester filler.

➤ Sanding down, either by hand or using a machine, until you get a smooth finish is one of the most gruelling of jobs, yet it is essential in terms of the final result. The fine dust from the filler gets everywhere, making eye and breathing protection practically useless. Up to six kilos of filler had been used to smooth these surfaces.

➤ After the first coat of primer and subsequent applications of filler followed by intermediate sanding, we painted the surface using a coat of grey sealant, again covering over the small holes and highlighting the imperfections that still needed to be sanded down. This colour contrast allowed for a better appreciation of the smaller imperfections.

➤ Using a pencil we then marked the remaining areas requiring work. These were then filled again and sanded down, for what seemed like the thousandth time.

➤ Little by little, however, the interiors started to improve. Before we got round to giving these surfaces a final coat we had spent fifty man-hours working on them.

➤ For the final finishing coats on the bulkheads we used a satin white. The first of these was applied with brush and roller, to ensure that the paint penetrated fully into the undercoat.

➤ After this first coat a number of small imperfections still needed to be filled, and of course sanded down yet again. The work was interminable!

➤ Finally, and in order to get a better finish, the last two coats were sprayed on. For the interiors of boats it is far more usual to use brush and/or roller, given that the fine spray gets everywhere and ruins your varnished surfaces, ceilings and upholstery. But at this point Samba was stripped of any decorative elements so we could basically do as we pleased. When working on interiors with a spray it is also essential to ensure as much ventilation as possible inside the boat. Even so the atmosphere becomes unbreathable. Even the most hardened professionals have to take frequent breaks to get a breath of fresh air.

➤ By now the change is notable. The sudden sense of spaciousness that resulted from the white paint was incredible, making the boat seem so much bigger and more roomy.

Painting bilges, lockers and the insides of cupboards

At the same time as we were painting the bulkheads we also cleaned out the bottom of the bilges, the lockers and what would be the interiors of the cupboards. In new boats these finishes are generally done using gelcoat, which is a very appropriate finish for clean surfaces, but one that will show serious adherence problems (cracks, flaking, etc.) wherever you find traces of glue, old paint, grease or salt residue.

Some people may also find it strange that we keep referring to the insides of the cupboards in a boat without any. Although we had not started with the carpentry work, we had a good idea where the furnishings that would make up Samba's new interior layout were going to go. Painting the backs of the future cupboards now would really simplify things later, avoiding having to paint through doors and in the restricted spaces of cupboards or lockers, with the added risk of smudging them with paint.

➤ Scraping, cleaning, degreasing and vacuuming up the dust is all part of the laborious preliminary work that has to be done before the bilges and the sides of the hull can finally be painted.

➤ The future forward cabin started out looking as dismal as this.

➤ A couple of coats of paint have gone a long way towards cheering it up. It is not a waste of time painting parts of the hull that are later going to be covered over and hidden from view. Apart from aspects such as aesthetic improvement and protecting the fibre, painting the bilges, locker bottoms and the insides of the cupboards will minimise all those bad smells that old boats inevitably produce (mould, damp, etc.) and which can make life on board so unpleasant.

➤ The sides of the hull in the saloon also got their coats of paint. When you use a brush it is easier to control how much paint you are putting on, making sure it penetrates into the fibre and reaches all the hidden nooks and crannies without any problems.

➤ Before painting the bilges, you have to scrape off, clean up and vacuum away as much of the old paint, salt residues, oxide and dirt as possible. In any case it is impossible to completely clean up the bilges of a boat that is over a quarter of a century old. This is why specific paints have been developed just for bilges that key firmly on and cover well, qualities that are (in this case) more important than the impeccable texture of the finish.

The keying of a paint is its capacity to firmly adhere to poorly prepared surfaces (previous coats of paint, dirt, dust, grease, etc.). Covering capacity is the paint's capacity to form a uniform coat of colour on complicated surfaces (stone, glass, cement, fibreglass, cloth, etc.).

➤ The change in the look of the bilges compared with the previous photo makes the need for any further comment superfluous.

➤ The aft locker has also been visibly improved by the new paintwork.

➤ And you can say the same about the forward cabin chests, where it would now be possible to stow whatever you want without any risk of it picking up those bad smells. In this area, as the fibreglass of the hull is flatter you can use a roller, which always speeds things up a bit.

➤ As an example of how much better things start to look with a lick of paint, just follow this sequence in the space under the sinks in the galley. They look so much better once they have been cleaned up and painted.

List of materials used	Quantity
Polyurethane Acrylic Paint 2 components	4.5 l.
Thinner	2 l.
Epoxy Filler 2 components	1.5 l.
Polyester Filler	6 kg.
Primer	5 l.
Bilge Paint	9 l.
Paint Stripper	0.75 l.
Primer 2 components	4 l.

➤ The quantity of materials needed to paint the interiors will obviously vary with each boat. Apart from differences in length or volume, the state of the surfaces being painted and/or what lies under them will also have a big influence on these needs.

Varnishing the mouldings

While they were doing the interior paintwork, the painters also started to renovate the varnish of the bulkhead mouldings with their attractive teak finishes, standard fittings for most boats more than a couple of decades old. These would be considered an authentic luxury in a modern production boat.

➤ Unless the old varnish is in perfect condition, in which case all you need to do is prepare it and give it a fresh coat, you have to start a complete re-varnishing by stripping it down to the bare wood. Over time varnishes tend to get darker, giving the wood a rather lugubrious quality that has nothing to do with the original hue. Start off with the orbital sander working on the most accessible areas. Then change to finer grain disks for the roto-orbital sander, and finally finish off the most difficult nooks and crannies using a fine-grain sanding sponge, or fine-grain paper. On Samba we did the sanding before the bulkheads had been given their final coat of paint, so as to avoid accidentally damaging the final paintwork finish.

➤ The mast step, consisting of two pieces of solid oak bolted onto both sides of the main bulkhead, had small round wooden inserts, the size of coins, crudely disguising the heads of the bolts. In their place we filled the gap with a paste that had the same tone as the oak itself. Once this had dried it was sanded down and varnished as if it were wood.

➤ Having eliminated all the rest of the varnish and prepared the mouldings, using fine-grain sandpaper or emery paper, we then started work on the varnishing itself. First a coat of sealant was applied to penetrate deep into the wood and serve as a support and protection for the successive coats. Some varnishes allow for a diluted coat of varnish to be used as a sealant. The important thing is that the first coat penetrates deeply into the pores of the wood.

➤ Unlike the sanding, the last coat of varnish was added after the bulkheads had been painted, which meant that we also had to clean up the occasional paint smudge on the mouldings. We gave the mouldings two coats of a double component satin-finish varnish. When working on the interior of a boat the use of this kind of varnish is not essential, you can use a single component varnish with no trouble at all, but we wanted to use the double component varnish here because the mouldings form part of the furnishings that suffer the most from scratches and scrapes. The highlighted natural tone of the teak, in contrast to the white paintwork gives a good idea of the interior style we were seeking for Samba.

Boat owners are often faced with the dilemma of whether to fix their boat's main engine or replace it altogether. If you wait until your engine has actually given up the ghost, then the decision will have been made for you. But as long as your old engine is still chugging along it can be a difficult decision to make.

Time for a change

How long can you realistically drag out the working life of an engine? Is it better to keep on repairing an old engine or replace it altogether? It is difficult, not to say impossible, to come up with a categorical answer to these kinds of questions, which we had to face up to while refitting *Samba*, in the same way as many other boats of her generation.

After a certain age (say 15/20 years) or a certain number of operational hours (say 1,500/2,000) it can be said that a sailboat's engine has paid for itself, although this does not mean that you have to replace it immediately. As long as the engine continues to

work you can delay any such decision, but you must also accept that your engine's useful working life is nearing its end and that now is not the time to invest in large-scale or expensive repairs.

Mechanical restoration is a basic step in the refitting of any boat but, before taking a look at the work we did on *Samba,* we would like to go over the reasons that led us to decide on a change of engine, as it is possible that our approach might be of use to others who find themselves in a similar position.

The original engine installed in *Samba* was a Perkins Marine 4.108, adapted by Solé Diesel. While the engine was, in broad terms, still working it was beginning to show signs of its age: erratic alternator, injection problems at start up, rusty pipes, dodgy looking exhaust pipes, untimely overheating problems, etc.

The first thing we did was contact the engine manufacturer and get their opinion. In a very professional way they warned us of the dangers of restoration, mainly due to the difficulty in finding spare parts for an engine that they had stopped making many years ago. Some dealers still keep a stock of these parts but tracking them down is becoming more complicated every day.

The manufacturer offered us a restoration service for their old engines, where they strip them down, repair them and leave them practically as new after passing through their workshops. However, they did not recommend this in the case of *Samba*, "What with stripping the engine down, repairing it and fitting it again the cost would be considerable, and you'd still have an old engine, with hardly any spare parts to be found, which could develop a fault anywhere, even after a thorough servicing", they informed us with commendable honesty.

I made the same enquiry of a mechanic who I trust, whose opinion turned out to be a practical version of the above, "With *Samba* you've got two options, the first is doing nothing at all, which means changing the oil, checking the belts, the alternator, the filters and stick with it until the engine has finally given up the ghost, which could happen tomorrow or perhaps three years from now. And the second option is to completely change the engine. Anything in between would just be throwing your money down the drain, trying to fix what can't be fixed. The bolts on old engines are always

➤ *The original engine fitted in* Samba *was a Perkins 4.108. Although this old engine was still working, signs of age were beginning to appear, unsurprising after over thirty years of faithful service.*

rusted solid and break when you try to remove them, and the same is true of just about any other part. In the end, trying to restore these engines is just a question of wasting a lot of time and hardly achieving anything".

On the basis of these opinions, and after a few days' deliberation, we decided to install a 55 HP Yanmar 4JH3E. Despite the increase in power this engine was actually smaller than the old Perkins, which meant that it was fairly easy to install.

Once we had taken the decision to replace the engine we then followed the steps indicated below. In the present chapter we shall be dealing with the preparatory work (removing the old engine and preparing for the new one), leaving the work of actually fitting the new engine for a later chapter.

Step by step

➤ *Having decided on a change, the mechanic started work on removing the old engine: removing the bolts from the engine mountings, uncoupling the shaft, disconnecting cables, fuel and exhaust pipes. All of which took just over an hour to do.*

➤ *Instead of taking the engine to a scrap yard the mechanic loaded it into his van and took it to the industrial college, where he had studied to be a mechanic. The lecturers at the college are always delighted to get their hands on these engines, which allow their students to get experience working on 'real' engines. In their practical workshop classes they usually only get the chance to work on new engines, which are provided by the manufacturers, good enough for teaching theory but, when it comes to the practical difficulties of working on a 'real' engine, there's nothing better than having a go at one that has put in many years of service.*

➤ *Finally, after trying out every possible angle of approach we managed to pull the engine out, upside down and lifted from the aft end. Even so it was a question of millimetres and we first had to disconnect the silent blocks and the oil filter.*

➤ *Along the way we ended up with this pile of peripheral engine fittings. Among these you can see fuel and exhaust pipes, the old instrument panel, soundproofing foam and the rust-covered manifolds. Somehow the fuel gauge, which was still working perfectly, also ended up being scrapped. Later on we had to buy a new one, exactly the same, as this model is the only one that can be connected up to the sensors in the fuel tanks. Another reminder of the adage that 'nothing must be thrown out that is not evidently useless or has not been satisfactorily replaced'.*

➤ *This is the depressing sight of Samba's engine compartment once the engine had been removed. However, by the time we had cleaned off all the flaking paint, given it a thorough degreasing and two coats of paint there was a vast improvement. Instead of painting the bottom, walls and ceiling of the housing we cleaned them up and prepared them for the installation of the new self-adhesive soundproof panels.*

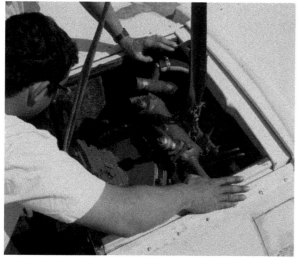

➤ *Removing an engine from its location on board is not as easy as it might sound. Many of these old engines were fitted before the deck had been fitted and the entrance to the cabins had not always been designed to allow for the removal of the engine. The mechanic explained to us that, when faced with the physical impossibility of removing an engine, on occasion he has even had to start breaking it down inside the boat, using a radial saw. The removal of peripheral fittings, such as cooling filters, manifolds, etc. usually has to be done in any case, in order to reduce the external dimensions of the block to allow for successful removal.*

➤ *Before fitting the new engine we had to restore the shaft, stern tube and propeller end bearing. The stern tube was the only one of the old parts that we were still be able to use, and the mechanic took it away to his workshop to be restored. In order to allow for the increase in power the shaft had to be remade in a larger diameter than the original. This meant that the propeller end bearing also had to be replaced.*

➤ The work done on the stern tube, including turning it to allow for a larger shaft diameter, was splendid, it looked like new! The stern tube was of the stuffing box type (a name that comes from the hemp fibre that used to be stuffed into the stern tube to avoid water getting in).

➤ These days hemp fibre or oakum has been replaced with lubricated synthetic fibres, graphite in our case. Graphite fibres are more resistant to use, offer greater watertightness and, as they cause less friction, do not heat the shaft up, thus avoiding long-term damage. As you can see in this photo, the stern tube includes its own pressurised greasing nipple.

➤ With the shaft provisionally in place, and using the dimensional layout parameters of the new Yanmar engine as a base, the mechanic then took the necessary measurements and angles for fitting the engine to the mountings. The dimensions of the Yanmar silent blocks, while not exactly the same as those of the old Perkins, were quite similar. This being the case we did not have to completely redo the mountings, it was easier to adapt the engine supports, adjusting them to the exact dimensions of the housing.

➤ The range of combinations of propeller end bearing measurements (internal and external) is wide, although not unlimited. In some cases, such as ours, sought after combinations do not exist and you have to 'adapt' a standard bearing. Given the evident difficulty involved in stripping down the propeller strut to grind it down, so that it will fit the bearing, and then laminating it back into place on the boat, it is easier to grind down the bearing to adapt it to the existing propeller strut. This is something the mechanic was able to do in his workshop.

➤ Finally, the shaft, the bearing, the stern tube and the bracing plate could all be fitted together. Once the engine's electrical installation had been redone we could then start to think about installing the new Yanmar engine in Samba.

Chapter 16

Bow pulpit

Restoring Samba's bow pulpit, a job that basically only required us to clean up and polish the old scars left by the railings and welds, was the perfect opportunity to give in to the temptation to modernise the line of her bow and introduce a frontal opening.

A touch of haute couture

If an amateur wants to improve his boat with any 'customised' part or adaptation in stainless steel, my first advice is that you find a good metalworker and, if possible, one with experience in the nautical field.

To start with, the work of restoring our bow pulpit only required us to clean up a few rusty welds and accidental scratches. Rust will always appear wherever stainless steel has been welded and the metal has not been correctly scoured. However, we had another problem with the pulpit, which was the misalignment of the fixings, which had come into contact with the gunwale and were being electrolytically attacked. Having decided to remove the pulpit so that it could be fixed, we thought we might as well modernise her lines by introducing a frontal opening.

The work of opening up a bow pulpit is a made-to-measure job and will be unique for every boat. There is no way you could find these parts ready-made or 'off the peg' in the kind of catalogues where you might find all your other stainless steel fittings, such as stern platform railings, stanchions or cockpit ladders. The cost of having a 'customised' stern platform made up could easily be three or four times the cost of fitting a production model. More than a difference in the quality of the materials or finishes, what you are paying for here is craftsmanship. It is the same as getting a suit made-to-measure compared to just buying one off the peg at your local department store.

Ninety per cent of the cost of transforming a pulpit, as illustrated in this chapter, goes on paying the craftsman. The raw material, 316L marine stainless steel tubing in this case, represents barely ten per cent of the final invoice. This work could have been done without removing the pulpit from the boat but then you would have to evaluate whether paying travel expenses and time, as well as the inconvenience of working on board (which represents a supplement in terms of hours) would compensate for the time spent removing the pulpit and fitting it again. In most cases bow pulpits are held in place by bolts that are easily accessible from the anchor locker.

➤ To begin with the only restoration work actually required on the bow pulpit was polishing the welds, many of which had rusted, and redoing the fixings, which had become misaligned and, having been in contact with the boat's gunwale, had suffered from electrolysis. In the end we could not resist the temptation to modernise her lines by introducing a frontal opening.

Step by step

➤ At the workshop, after measuring the angles and length of the new parts to be welded on, the welder bent the lengths of stainless steel tubing using a hydraulic machine specifically designed to do without the need to heat it.

➤ Once the welder had made the cuts using a radial saw it was too late to change our mind.

➤ Once bent the lengths of tubing were lined up so that the position of the cuts could be marked, verifying the symmetry of the pieces and their adaptation to the new pulpit design. At the same time two crosspieces were tack-welded to the pulpit to avoid it becoming misaligned when cut.

➤ This is the part of the original pulpit that we had cut off. In the scrap skip at the entrance to the company there were various similar pieces, indicating that we were not the first to have our pulpit adapted and opened up in this way.

➤ The next step was to weld the new curved sections of tubing into place. In order to simplify alignment, and provide greater rigidity, small lengths of lower gauge stainless steel tubing were inserted inside the ends of the parts to be joined.

➤ Once the lengths of tubing had been aligned and fitted into place they were spot-welded to each other.

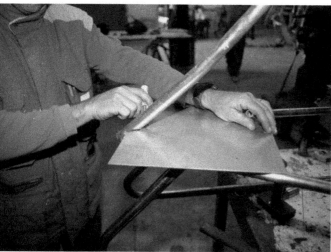

➤ At the same time as the work with the tubing was being done, the shape of the stainless steel plate, which would serve as a boarding support step, was also traced out. A piece of teak would later be affixed to this plate.

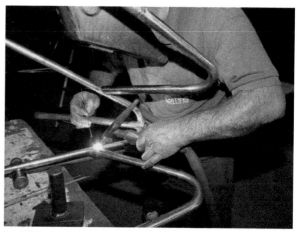

➤ With the pulpit on the work bench, the operator completed the welding of the old lengths of tubing to the new ones, an operation that left a fine bead around the joint. This work could also be done with the pulpit in place on the boat, but there are always complications with welding, and the subsequent polishing work, and this would tend to increase the final price.

➤ Once the support plate had been cut to size it was also welded into place and, when this had been done, most of the work had been completed and you can make out the new pulpit design. All that we had to do now was the polishing and finishing work.

➤ The first stage polishing was done using a rotary grinder and rough grain sanding disk (60/80), to eliminate the weld beads, such as the one you can see in the foreground here.

➤ The next step, after this first polish is very important, though sometimes forgotten, and involves covering all the welds with a specific pickling paste to eliminate the iron that has oozed out on the surface of the stainless steel as a result of the heat of the welding. If you do not do this, the rust will be back again in the medium term, spoiling the appearance of the whole assembly. After a few minutes the pickling paste is washed off using fresh water.

➤ Little by little the polishing is done using finer grade sandpaper (150/220), until the outlines of the welds are left completely smooth. In the end the joint marks will disappear and it will be impossible to see where the lengths of pipe were originally welded.

➤ The last buffing is done using a disk of special wool to apply the brilliant finish that you always want for a boat's stainless steel fittings.

➤ The rejuvenated look of the new bow pulpit was really attractive. All we needed to do now was redo the fixings that attach it to the deck. Apart from the teak step/seat, we had also fitted a socket for the gangway. This ingenious fitting, which is mass produced, will adapt to most bow pulpits, simplifying access whenever your boat has been moored at the bow.

➤ In order to redo the fixings it is imperative to do the work on board. First the mounting plates are bolted into place, perfectly aligned to the gunwale, and then the feet of the pulpit are spot-welded, to make sure that they will remain in position. Now all that you need to do is remove the pulpit again and take it back to the workshop so that the mounting plates can be firmly welded in place, polished smooth and buffed to a brilliant finish, just like the rest of the assembly.

➤ The new pulpit has given Samba's bow a more modern air, maintaining her overall lines while notably improving the finish and, above all, the functionality of the pulpit as a whole.

We started the work of preparing Samba's head for her new furnishings by lining the walls with synthetic laminated panels.

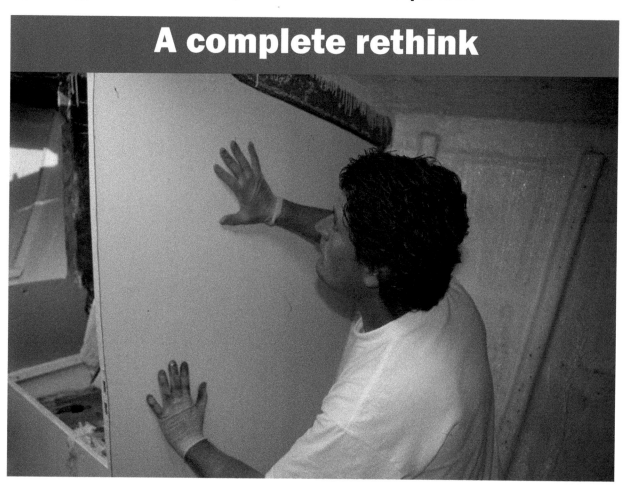

A complete rethink

The original layout of the boat included as many as 11 berths, four of which were located in her central cabin. As the years went by, these berths were replaced by a makeshift head, the finishing of which left quite a lot to be desired. While part of the old structure still remained, a washbasin and a toilet were installed here. The cabin as it was, apart from being ugly and uncomfortable, was not fitted with any cupboards or lockers. The walls, lined with wooden planking of different shapes and colours, were also a problem, giving this space a completely shambolic appearance.

We started the work of refitting this central cabin space, converting it into a real head, by ripping out all traces of the original finish. The next step was to fill the holes left in the forward bulkhead by the original berths. This had previously been done in a very rough and ready way, simply by screwing on wooden panels to cover them over.

The mosaic of finishes and materials exposed on the bulkheads on either side of the cabin, following this work, obliged us to come up with an effective lining system. In view of the irregularity and roughness of the surfaces it would have been exceedingly difficult to smooth over the bulkheads, allowing them to be either painted or lined with some light finish (carpet, vinyl, etc.). The solution that we finally came up with was to use panels with a synthetic finish.

Everybody knows this type of synthetic finish (Formica-type laminate), having seen it on endless furnishings and cladding finishes, so there is little new that I could tell you about its hard-wearing and resistant qualities. The standard version sheets of this lining are very fine (0.7 mm) and can be glued onto wood or completely smooth surfaces, providing a perfect finish. The problem arises when they are used to cover convex areas or bumpy and irregular surfaces because the sheets tend to mould themselves to the underlying form, resulting in a poor finish with a lumpy appearance.

Fortunately you can get hold of thicker sheeting (anything up to 6 mm) specifically to avoid this problem. In our particular case we decided on 3 mm thick sheeting, with a white, satin finish, the colour we liked best. The catalogue also included imitations of various types of wood, a solution increasingly sought after for the linings (floors, bulkheads, etc.) used on large production boats.

Step by step

➤ As mentioned before, Samba's original layout included up to eleven berths, four of them in her central cabin. As time passed some of the berths were replaced by a head and shower cabin. This so-called improvement, however, still maintained part of the structure of the original layout, adding a toilet, which was previously located in the forward cabin. Apart from being ugly and uncomfortable, the head had no cupboards or lockers.

➤ The restoration of Samba's central cabin, converting it into a real head, started with stripping out the original finishes. This work was done by the owner on his own, and it took him a full weekend.

➤ Another problem was the aesthetics of the wall surfaces, with strips and sheets of wood glued on in what seemed like a completely random manner, giving this cabin a shambolic look.

➤ Stripping out these finishes revealed holes in the bulkhead, left by the original berths, which had been covered over in a rather rudimentary fashion using sheets of wood screwed to the bulkhead.

➤ The work of filling these holes took the carpenter about a day to do. First he made a template of the shape of each hole and then cut out sections to measure from a 25-mm thick sheet of marine plywood. The pieces were then fitted and glued into place.

➤ The outlines of the joints were then covered with three layers of fibreglass to reinforce them. Structural solidity is one of the best known qualities of the North Wind 40 series, but even so bulkheads will always flutter lightly and imperceptibly at sea (in many boats these movements are much more evident). This means that the mechanical strength of a fibreglass laminate is important in terms of avoiding the eventual formation of cracks in the glue that holds these sections in place.

➤ Having filled the holes left by the berths, permanently this time, we used a rotary sander to sand down the most visible lumps and irregularities in the fibreglass, at the same time smoothing down and preparing the surface of the bulkhead as a whole. This left it with a matt finish that would provide a key for the polyurethane adhesive that we would be using to glue on the panels.

➤ As was the case in other parts of the boat, and after thoroughly eliminating the dust and dirt caused by the previous work, we continued the restoration work by applying two good coats of paint to the inside of the hull, which would be the back of the future cupboards in this cabin. It is much cleaner and easier to do this before the cupboards and other furnishings have actually been installed.

➤ Once the inside of the hull had been painted, and with the bulkheads cleaned, degreased and smoothed down as much as possible, the time had finally arrived to glue the sections of Formica sheeting into place. First of all, to trace the outline of the four sections, we used large sheets of paper to make up templates.

➤ These templates were then taken onto dry land and laid on top of the Formica sheets. The next step was to trace out the profile of each section. The silliest mistake you can make at this stage is to lay your template over the sheet the wrong way round and trace out the 'mirror image' of the piece you actually want.

➤ Using a jigsaw you can then easily cut out the outlines. Before gluing the pieces into place you must first check that the fit is precise. You will inevitably find that some corner or edge needs to be filed down or adjusted slightly.

➤ Once you have made the finer adjustments to the four pieces, and having once more thoroughly cleaned the surfaces of the bulkhead, you can apply the polyurethane adhesive to the surfaces to be glued. This type of adhesive combines excellent adherence with the necessary elasticity required when lining bulkheads. You must remember that all bulkheads inevitably suffer imperceptible movements when sailing.

➤ After applying the adhesive, each of the pieces of Formica sheeting must then be carefully fitted into place. To avoid the sheets from springing loose before the polyurethane glue has had time to set, we inserted small shims (spacers) at the bottom.

➤ It is also important to apply pressure evenly over the whole surface of the Formica sheet sections. To do this we used strips of wood jammed between the two surfaces, simultaneously pushing against both sides. Parts that are glued together using polyurethane adhesive, unlike the case of traditional glues, should not be pressed too firmly. To work correctly this material requires a certain thickness, depending on where it is used, sufficient body to ensure a strong yet slightly flexible union.

➤ This is what the head looked like at the end of this first stage. We have stripped out the old fittings and furnishings down to a solid base from which to start the restoration work. The white walls give an agreeable sensation of space and will ensure ease of maintenance required in such a damp area. In the photo you can see the electrical installation wiring. As we have mentioned in the chapters concerning this aspect, work on the electrical installation continues in parallel with the rest of the interior refit.

Painting locker covers and berth bases

In a refit such as the one that *Samba* underwent, you should not forget the berth bases and locker covers. Some of these still had the original old paint finish, badly discoloured after so many years, others had been repainted at some time or other in a variety of colours and others had no protection at all. We took advantage of the occasion to make ventilation holes in the bases and covers, something that has become standard for some production boats.

➤ Although, in our case, this work was done by a carpenter, in fact it is not particularly difficult and a perfect opportunity for anyone who's a dab hand at do-it-yourself to do it themselves. The items will be hidden from view anyway and any mistakes made can easily be sorted out. You can use just about any paint or varnish without problems and apply it using a brush, roller or spray. And if you make a serious mistake you can easily go back and start again or even make up a completely new cover or base.

➤ It is always a good idea to remove from the boat any parts that you need to paint or varnish. It is difficult to believe how much space the berth bases and locker covers occupy when piled up on dry land.

➤ Before repainting over old layers of paint or varnish you should prepare them as well as possible, to improve the adhesion of the new coats. We did this work using a rotary sander. We also took advantage of the occasion to make ventilation holes, which avoid the formation of mould and bad smells in the lockers. An electric drill fitted with a wood boring spade (on sale at any hardware store) is the perfect tool for this job. To ensure interior ventilation without weakening the resistance of the panels or compromising the comfort of the mattresses or cushions, the holes should be 3 to 5 cm in diameter and the separation between each one and the sides should be at least 1.5 times their diameter.

➤ The covers and bases are later given a first coat of primer followed by two coats of satin-finish white paint. Applied using a spray gun, the successive coats of paint breathe new life into covers and bases while providing them with the necessary protection.

➤ In order to obtain a good finish and to make sure that the paint adheres correctly, it is important to eliminate all the dust left by the sanding of the surfaces. The best way of doing this is to blow it away with compressed air, although you can also use a brush or a cotton rag soaked in turps.

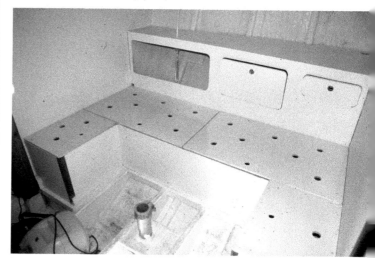

➤ Although the locker covers and berth bases are generally hidden from view under their respective mattresses and cushions, it is always nice to know that they look good enough not to be embarrassed by a 'Full Monty'.

Chapter 18

After several months in dry dock Samba finally dried out and her hull was at last ready to receive its epoxy treatment. This operation was to protect the hull from osmosis over many years.

A new epoxy skin

In chapter 4 we showed how *Samba*'s hull and keel were blasted with a mixture of sand and water to eliminate the gelcoat, which was in a bad condition. Since then, and once the high level of hull humidity had been checked over, the drying out process began, in the style of the best cured Serrano ham. In order to assist with this task, and although it may seem contradictory, we washed the hull down at regular intervals using fresh water. This helped to eliminate the salt and acids encrusted in the pores of the fibreglass.

It is impossible to say exactly how long it will take a hull to dry out. This will depend on how damp it is, the climate and many other variable factors. Generally speaking somewhere between two and four months is enough if you have adequately purged the fibre.

As soon as *Samba*'s hull was as dry as parchment, the moment had arrived to replace the external 'skin'. Numerous opinions exist regarding the best treatment for the prevention of osmosis. Each shipyard and brand of hull paint will swear that they know best. More than judging the differences, what we tried to do was compile a synthesis of the points where all of the different brands and application systems agreed.

If there is no definitive consensus regarding the best method of purging the damaged fibre (basically peeling or sand blasting) there is one thing that everybody seems to agree about and that is that the best epoxy treatment will serve no purpose if the hull has not been previously dried out to acceptable levels. Only if this is the case can a satisfactory result be achieved. If this is not the case, the humidity encapsulated in the fibre will continue to attack the laminates.

Another aspect on which everyone seems to agree is the use of resins, fillers and epoxy primers for the treatment of osmosis. Without getting too technical here we can say that epoxy resin is the 'rich relation' of polyester resin, the massive use of which in boat building is directly associated with its price, which is far lower than that of epoxy resin. On the other hand, epoxy is much harder and more impermeable than gelcoat (derived from polyester), and is the most frequently used material in the treatment of osmosis. A good epoxy treatment can guarantee ten, fifteen or even twenty years of protection against water infiltration.

Essentially all epoxy resins, fillers and primers have similar chemical structures. With a view to osmosis protection treatments, the most practical are the ones that will allow for a good thickness in each coat and good adhesion of the application. But, in the end, the most important thing is to follow the usage instructions and the curing times for each brand and product.

The subject of osmosis prevention treatments and

applications leaves little room for amateur involvement. These are long and laborious processes, in which only the experience of a good professional will ensure a good result.

Finally we must also stress that osmosis is a completely normal, even an inevitable, problem for any boat with a fibreglass hull. There is no need to be alarmed if you discover that, after a few years, your hull is starting to absorb water because modern treatments can resolve the problem simply and satisfactorily. In the same way that sails, engine or rigging will all have to be replaced during the useful life of your boat, the exterior 'skin' of the hull will also have to be restored every ten or twenty years.

➤ Epoxy-based products are expensive but, in the end, they are only going to represent about 15% of your final bill for osmosis treatment. The lion's share will go on the man-hours that have to be dedicated to this laborious process.

Step by step

➤ Treatments to cure or prevent osmosis have to be applied to a dry hull lacking internal humidity. The photo on the left shows the humidity measuring gauge after Samba's hull had been peeled. You can see how the needle is completely off the red end of the scale (indicating maximum humidity). In the photo on the right we see the same reading taken several months later, with the needle barely entering the green area (indicating minimum humidity). It is only when you have achieved the latter situation that preventive or curative treatments can be considered.

➤ Although osmosis only affects the fibreglass on Samba's hull we took advantage of the occasion to apply the same protective treatment to the keel, in order to avoid the formation of oxide.

➤ Treatment started with sandblasting the whole of the keel in order to completely eliminate rust, a job that the operator did hidden away behind these plastic curtains to avoid sand blowing about and getting into every boat in the area.

➤ *Immediately afterwards we applied the first coat of epoxy resin to seal the pores in the metal. It is important to do this as quickly as possible because the humidity in the air acts very quickly and will contaminate metal in no time at all. The best approach is to blast one side of the keel and then seal it before starting work on the other side.*

➤ As with the keel we also applied a coat of epoxy resin to the hull, now clean and dry. The purpose of this first coat is to seal the small pores in the fibre.

➤ Once the resin had dried we then prepared the epoxy filler, blending its two components (filler and catalyst) until we achieved a perfectly uniform paste.

➤ The first coat of epoxy filler, apart from its protective function, is to smooth over the most obvious irregularities in the fibre, and also in the casting. The whole surface has to end up completely smooth.

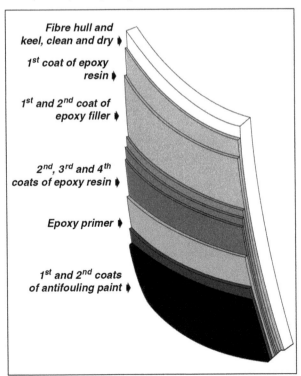

Fibre hull and keel, clean and dry ▶

1st coat of epoxy resin ▶

1st and 2nd coat of epoxy filler ▶

2nd, 3rd and 4th coats of epoxy resin ▶

Epoxy primer ▶

1st and 2nd coats of antifouling paint ▶

➤ Summary of the anti-osmosis treatment applied to Samba.

➤ Little by little, keel and hull receive two coats of filler. It is not easy to apply the filler in a way that forms a sufficiently thick coat and, at the same time, one that is both smooth and uniform. If you apply it too thick or too irregularly then a lot of time will have to be spent sanding the hull down. If you put on too little then you will have to repeat the operation with successive coats wherever required. It is hard to believe how big a 40-footer's hull is when you have to sand it down or fill it. The working position is often uncomfortable and physically exhausting.

➤ Between one coat of filler and the next, the hull had to be sanded down in order to smooth out the imperfections and, at the same time, open up the pores so that the previous coat would provide good adhesion to the following one in the next stage of application.

➤ The treatment of the hull continued with three coats of epoxy resin, the purpose of which was to complete the sealing and increase the impermeability of the hull.

➤ We then applied a final coat of primer, also epoxy-based, as a substrate for the good adhesion of the antifouling paint. Pure epoxy resin, as with the majority of this material's derivatives, crystallises on the surface as it dries, preventing effective adhesion of subsequent coats of paint, varnish or antifouling. The coat of primer is a way of avoiding having to sand down the whole of the hull, opening up pores in the epoxy, every time that the coat of antifouling paint is renewed.

➤ The antifouling paint that we applied can be left for up to six months without entering the water. But, because there were a number of jobs still pending we only applied the first coat. We would be applying the second, and last, coat when we are ready to relaunch her. Initially it might appear more logical to apply both coats when Samba was ready to be relaunched, but this solution would have left the epoxy primer exposed to the weather. This material, despite its powerful virtues in all forms (varnishes, resins, primers, etc), is rather vulnerable when it comes to ultraviolet rays. This means that it is not a good idea to leave it as the surface finish coat.

There are few things that will improve the looks of a boat more than painting her topside. But to do so correctly the work must first be meticulously prepared. You will have dozens of hours of sanding to look forward to, the application of several coats of primer and the interminable intermediate work of filling and sanding again.

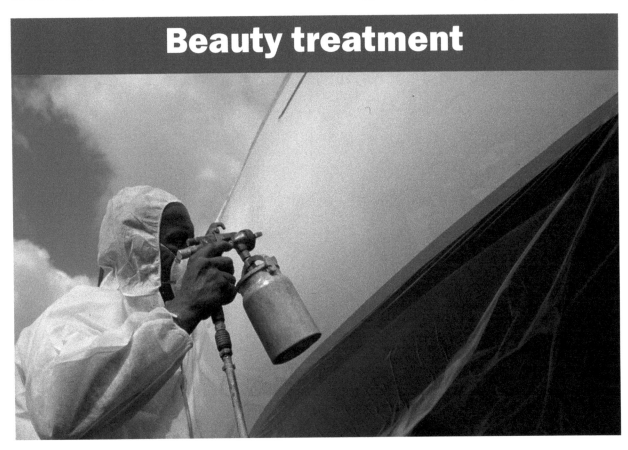

Beauty treatment

We gave *Samba* the osmosis treatment and, almost simultaneously, a complete paint job. This allowed us to take better advantage of the time, a lot of time, by alternating work on the keel, the hull and the deck. Preparing a hull for painting is meticulous and painstaking work. Numerous steps, all of which must be treated seriously, lie between the initial surface sanding and the last coat of paint.

An error often made by amateurs is the belief that a finishing coat is going to disguise small surface defects. In fact the opposite is true. Successive coats will reproduce, or even exaggerate any pore, crack or fissure, spoiling the final result. To achieve a perfect finish you will have to be meticulously careful, making slow but steady progress and steering clear of shortcuts.

Samba's main problem, apart from the kind of knocks and scrapes inevitably suffered by any boat, was that her gelcoat, dried out and mistreated over the years, flaked off in great quantities every time we set about sanding her, something that tends not to happen with newer boats. In any case, there is no way you should expect to prepare the topsides of a hull for painting any faster than 1 square metre per hour.

Amateur or professional painting

The paint manufacturers' leaflets will tell you that many of their products can be applied using a brush or a roller and by amateurs. While this may be true, the results are never going to be as spectacular as a spray gun in the hands of a professional.

If an amateur wants to collaborate in the painting of his boat, with the aim of saving money and/or gaining experience, the best thing he can do is start at the bottom, with the sanding and filling. This is the best way of shaving hours off the bill, even if it is at the cost of eating a lot of dust and wearing yourself out by the end of every day's work.

You will find that the time a professional spray-painter needs to give the topside two coats of paint will be somewhere in the region of 2% of your final bill. The figure is almost ridiculous in comparison with the importance of these two coats in the final result. It would be a shame to spoil all of that preparation work through sloppy finishing.

When painting a boat, the most important part of the bill (in terms of remunerated hours) is the man-hours spent preparing the surface for painting, Materials scarcely represent 10/15% of the total. This means that investing in quality paints is not going to make a great deal of difference to your budget. You must not forget, however, that the quality of the preparation and painting work is even more important than ensuring that you use top of the range materials.

When working in the open air it is advisable not to paint when it is either too hot or too cold. The condensation of water resulting from a change in temperature and/or humidity, particularly at the beginning or end of the day, will form small drops of water on fresh paint that will leave ugly blemishes.

Applying too much or too little paint is also to be avoided. And then there is the wind problem, responsible for many an open-air painting disaster. If you are in a hurry to get the painting done by a certain date this may pressurise you into choosing a windy day. Ignore this pressure, dust and dirt will simply become ingrained into your recently painted hull.

Finally you must also remember that, when using a spray gun, you also have to make sure that any other boats on the leeward side (cars too for that matter) are protected. If not, clouds of paint can easily be wafted over, leaving your neighbours with a permanent reminder of your ineptitude.

Step by step

➤ Samba *was just begging for a bit of beauty treatment. The gelcoat was peeling off and the whole of her hull was mottled with matt patches. The ageing process could also be seen in the multiple scrapes and scratches left by almost a quarter of a century of charter work.*

➤ *The list of materials that we needed started with vast quantities of different grades of sandpaper. There were also solvents (5 litres), degreaser (3 litres), brushes, rollers, masks and gloves, epoxy filler (a total of 5 kg), polyurethane primer (10 litres) and two-component polyurethane paint (12 litres).*

➤ *The first stage was to sand down the whole of the hull using relatively coarse-grain sandpaper (220) to prepare the gelcoat, cleaning it of impurities and providing a good adhesive surface for subsequent coats. On old boats, this first sanding will also loosen innumerable flakes of old, peeling gelcoat, which will require hours of filling and sanding to repair.*

➤ *Next the painters started to fill and sand down the most obvious knocks and cracks. In the photo you can see one of the painters applying polyester filler to the topside while another is sanding down the epoxy filler that we used as part of the anti-osmosis treatment on her hull.*

➤ After filling the most obvious imperfections a second complete hull sanding, using finer grain sandpaper (300) little by little starts to smooth out the irregularities of Samba's topside. During the painting preparation process for a boat you should not expect to advance any faster than 1 square metre per hour, it is a really laborious job.

➤ The time has arrived to apply the first coat of primer. Apart from improving the base for subsequent coats, the primer also serves to penetrate by capillary action, and fill the pores and smaller cracks practically invisible to the naked eye, that can always be found in the gelcoat. If this was not done the air contained in these pores would later emerge at the surface, forming small but ugly craters in the paint.

➤ The preparation work continues, now using fine-grain sandpaper (up to 400) to sand down and prepare the primer. Once again this will expose the defects that need to be filled and primed again. In some areas we had to repeat this process as many as four times.

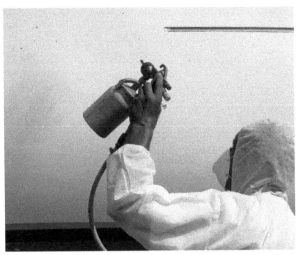

➤ Finally the finishing coat of paint can be applied; in our case we used a two-component polyurethane paint. For Samba, we applied the two required finishing coats needed to achieve consistent brilliance and colour. Before painting the hull we cleaned it again and meticulously degreased it. You must completely remove the greasy traces and impurities left by the sanding.

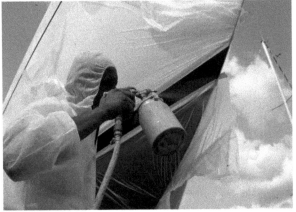

➤ Now that the hull is painted the time is right to trace out and paint the decorative blue stripes along the waterline and the top of the hull.

➤ Samba now looks really sensational. As we mentioned at the start of this chapter, there are few things guaranteed to improve a boat's elegance and poise like a good coat of paint.

Chapter 20

Painting the deck

Painting the deck brought to a close one of the most striking stages of Samba's refit. Without wishing to count our chickens, we could almost say that the painting of her deck heralded the turning point of Samba's renovation.

The icing on the cake

In previous chapters we showed you how we went about stripping all the fittings that were no longer needed from *Samba*'s deck. We completed preparatory fibreglass work for attaching the new rigging to the deck and repaired parts of the boat that were damaged. Now it was a question of covering over the traces left by all this preparatory work so that the refit could go ahead.

To simplify the painting of the deck we even removed the few fittings that were not due for replacement and re-installed them when we had finished. We could have masked these but, the more we cleared the deck, the

easier the work and the more splendid the finish would be. In the case of *Samba* we also had the advantage that her interior was stripped of any cladding or lining, revealing all the bolt and screw heads on the underside of the cabin roof. Had this not been the case the work of removing the fittings would have been much more complicated and harder than it was.

The passage of time, represented by the prolonged action of sunlight, the continuous attrition of salt and the wear and tear from the endless toing and froing of her crews had resulted in endless pores opening up and widespread peeling of her gelcoat. All these imperfections needed repairing before we could even start to think about repainting.

The list of materials used, as well as the process followed for painting *Samba*'s deck was, in general terms, similar to the work already done on her topside. It was a case of filling in chips and irregularities with polyester paste and then sanding it down, either manually or using a sander, and using sandpaper of increasingly finer grades. After that a coat of primer had to be applied and finally a top coat of two-component polyurethane paint.

There is one factor, that we might call 'perfectionism', which is always involved in the preparation of the hull, topside or deck. When you start to repair small defects, little by little, other lesser defects, which had not previously been noticed, begin to become evident. Professional painters, and amateurs too, will find themselves endlessly filling, sanding and priming, and then endlessly finding new places that need to be filled,

sanded and primed. The older the boat, the longer this will take. The key question is knowing when enough is enough.

Fortunately, or unfortunately depending on your degree of perfectionism, the day will come when you have to say, "That's it!" and decide that the time has come to get out the paint; even though you know that there are minute unfilled pores that are bound to show through or that a given area of the deck has not quite regained the glazed texture that you are after... In the end you just have to be pragmatic. If a detailed examination of a brand new boat inevitably reveals a number of small defects in the mouldings and finishes of her deck, why should your boat, with a quarter of a century of sailing behind her, be any better? Paradoxically, accepting this premise, a fine line between defeatism and pragmatics, implies setting your sites on the best feasible goal. Without going through this process, what in the end may be small imperfections would have ended up being major blemishes.

Step by step

➤ Filling small chips or holes on the deck gets to be a pretty repetitive job. First you have to clean up the hole using a sharp-pointed knife or chisel to eliminate the flaking gelcoat along with any exposed fibres. The edges of the hole then have to be cleaned up as much as possible to simplify the application of filler or polyester.

➤ Next you have to eliminate any traces of grease or dirt using acetone or any other strong degreasing agent. This is another necessary step to ensure the correct adherence of the filler.

➤ One of the main difficulties with filler is how much to apply. If you use too much then you will need more time and work to sand it down and if you use too little then you are going to have to give it another coat, to guarantee a smooth surface.

➤ You can work on some areas of the deck using a sanding machine but for others, particularly the interior corners of the cockpit, you will have to do it all manually, which often means adopting very uncomfortable positions.

➤ Finally, the deck was ready to receive its first full coat of primer. The absence of fittings and accessories made the work a lot easier for the painters, no need to mask all those fixtures and fittings. In the case of Samba all we had to do was cover up the bow pulpit, stern pulpit and the trim.

➤ Pre-painting preparation is extremely laborious, not to say a pain in the neck. In the case of Samba we invested 40 hours just in getting her deck ready to be painted. Again it is a good idea to alternate different stages of the work. While one person is opening up and cleaning holes, another can follow on behind filling them and later, while one does the sanding another can apply more filler, until the necessary smoothness of finish is finally achieved. On the sides of the cabin, for example, where the spaces occupied by the old Perspex panels had been laminated with fibreglass, we had to apply three coats of filler, sanding down each one, to get a smooth surface. And all of that after doing everything possible to ensure a near impeccable finish to start with.

➤ When repairing non-slip surfaces on the deck, the state-of-the-art solution is to reproduce the non-slip pattern effect before doing the painting. To do this you will have to make up moulds of the non-slip pattern using filler and then fill them with fibreglass laminates which can later be adapted to the area in need of repair. This process takes a long time and, in the case of Samba, we finally decided on a cheaper shortcut. This was to level out the repaired areas, until they could not be distinguished from the adjacent patterned finish and then, once the deck had been painted and the final coat of non-slip paint applied, the matt tone of the non-slip paint served to pretty much disguise the irregularities.

➤ The primer penetrated and filled the almost invisible pores of the gelcoat, sealing them and preventing the air contained in them from escaping later and forming small craters. In the flatter and more 'perforated' areas we applied a coat of high density primer, which served as a very good filler and covered hundreds of small pores that it would have been impossible to fill one at a time. In the areas where the non-slip finish had flaked off we applied the primer using a brush, individually filling the pores in the gelcoat. It's really difficult to fill imperfections where you have a diamond-shaped non-slip pattern. If you apply paste, when this has dried it is also difficult to sand it down between the embossed diamond shapes and you will end up with slabs of filler encrusted on the deck, an eternal eyesore.

➤ We thought the moment would never arrive but, finally, the deck was ready for the last two coats of paint, a job that, paradoxically, would only take a couple of hours to do. Before applying the paint, the deck had to be degreased. For this finishing coat we decided to use a white polyurethane paint with a very slight blue tint (the same as on the hull). What we were aiming for here was to achieve the shade of an older generation of boats, instead of going for a more modern 'nuclear white', just a question of taste. There are, however, a number of particular technical problems involved in painting a deck. One is that the guy doing the painting is going to be standing on the surface that he is painting, which means he has to avoid painting himself into a corner. A certain level of experience is necessary in order not to overlook any part of the deck while remembering not to stand on areas that you have already painted.

➤ Even after we had applied the two coats of polyurethane paint, the work was still not finished; we also had to repaint the non-slip areas. We masked these areas and, using a roller, applied a new layer of polyurethane paint. When a deck is repainted, the combined thickness of the primer and the top coats tends to slightly even out the form of the diamond-shaped non-slip pattern, reducing its functional effectiveness. If we had not added a final coat of anti-slip paint, the original effectiveness of these areas would have been significantly reduced, in the dry and the wet.

➤ For the non-slip finish we used minute ground-glass spheres, mixed in with the paint, which provide a slightly rough, matt finish. The moulded strips on the cockpit floor also have to be masked before receiving two coats of this preparation. Another alternative would have been to buy one of the special paints already prepared for this purpose. Some of these incorporate the very same micro-spheres that we added, while others incorporate other synthetic non-slip materials.

➤ Samba's recently painted deck was now impeccable; she looked as if she has just been launched! As we said at the start of this chapter, you could almost claim that, once her deck was painted, she had finally arrived at the turning point in her resurrection, at least on the outside.

Chapter 21

Deck hardware (part 1)

After so much time dedicated to the dirty work of stripping down, preparing and repairing the different parts of the boat, it was a real delight to see the shiny, brand new hardware and fittings, all piled up on Samba's deck. What a difference! All these brand new items to fit and no need to worry about dust, grime or bad smells.

Putting her back together

A glance at the above photo may give the impression that we just went out on a spending spree and bought half the shop. Not the case at all. Replacing deck hardware requires a great deal of thought. Before going ahead with the kind of preparatory work discussed in the previous chapters we had already decided on the new rigging set up that we would be installing in the future and also the different parts and fittings that you can see laid out in the photo.

This strategy, which we would recommend for any refit, was initially based on a keen observation of boats, both new and old, with similarities to *Samba* in terms of size and sail profile. That was where we picked up our first ideas. The rigging of modern yachts, for example, may be somewhat miserly in terms of size and/or number of elements but there is always a marvellous inventiveness in engineering and operation terms. Many modern solutions can also be applied to older boats. There is also a lot to be learnt from the catalogues and websites of deck hardware manufacturers, who usually set aside a number of pages for graphics and practical examples as a guide to purchasers.

In our case we mainly resorted to Lewmar parts and fittings to equip *Samba*'s deck. Although my purpose here is not to further enhance this firm's well-deserved reputation, the advantages of negotiating a large order with a single supplier often allows for a more advantageous overall price, while also guaranteeing a certain aesthetic and functional standardisation. The people at Lewmar also helped us with the specification of our needs, an area in which a bit of specialist advice is always more than welcome.

The point I would really like to get across is that all the work described in the previous chapters forms part of a greater scheme, in which all of *Samba*'s new rigging elements had already been defined. Any other approach is bound to end in trouble.

Almost all of the deck hardware is attached using four bolts and a bit of sealant. Although on *Samba* this work was done by professionals, most of it is well within the reach of the average amateur. All you need is a measure of common sense, another measure of care and attention and a few basic tools. Generally speaking when it comes to assembling winches, rope clutches or portlights there is hardly any difference between one make and another or one boat and another.

Step by step

Installing the hatches

➤ Installing deck hardware is well within the reach of an enthusiast of average skills and experience and, generally speaking, installing a hatch or mounting a winch is pretty much the same from one boat to another.

➤ When installing a hatch, the first step is to line up the lower frame on the prepared base area and clamp it in place. You must carefully square up all the sides and corners because any misalignment at this stage will have serious implications for the later stages.

➤ You also have to make sure that your hatch is correctly adjusted from inside the boat. This is the perfect moment to double-check and make sure that there are no impediments for any of the drill holes that you will be making, and also that you will have easy access when it comes to tightening the bolts.

➤ With the lower frame of the hatch held firmly in position and using one of the holes in it as a template, mark out where you are going to drill the first new bolt hole. When this has been done remove the hatch and drill this hole clean through the base. Fit the hatch back in place again and insert and tighten this first bolt. Now, carefully lining up the hatch again, mark out a second hole and drill it through, if possible diagonally opposite the first, and then insert and tighten the second bolt. When the first two (or three) bolts have been done up, make the rest of the holes and then insert the corresponding bolts. You should never do up all the holes one after the other without inserting and tightening at least the first two bolts. The risk of the frame shifting ever so slightly is enormous.

➤ With the hatch now in place and all the bolts inserted, the time has come to tape the outline of the frame to avoid the sealant making a mess of the freshly painted deck. Before doing this we countersunk the holes, to allow space for the sealant, forming a kind of O-ring, to prevent water seeping in.

➤ Now you remove the frame again and apply sealant to the whole perimeter of the base. Monocomponent polyurethane has great adhesive power, to the point that we have seen boats where the hatch frames are held to the deck by nothing more than this adhesive on its own.

➤ The next step is to once again fit the hatch in place and then finally tighten the bolts. If you want to fit the lower frame of the hatch precisely in position without any risk of smudging the sealant, one trick is to insert two small screwdrivers through diagonally opposite holes in the lower frame and then lower this down until the shanks of the screwdrivers slide into the respective holes in the base. This will also stop the frame from sliding about as you tighten up the first few bolts.

➤ The fore end of this hatch was right above one of the bulkheads; as a result we had to use self-tapping screws instead of the usual bolts.

➤ When tightening the nuts it is important to do so progressively working your way around the whole frame. This will conserve the uniform thickness of the sealant and avoid misshaping the frame. The final tweak must be hard and firm but never strenuous. Always stop before the frame starts to give.

➤ The next day you can eliminate the sealant that will have been squeezed out by the lower frame. To do so use a chisel made up from a piece of Perspex. Perspex is not as hard as steel, which means that you will avoid marking the aluminium or the paintwork below the tape.

➤ It took us about three hours' work to fit the three hatches. A small price to pay for such a stunning change.

Installing the opening portlights

➤ To line the portlights up in their correct position we used the inner frame of the portlight, which is lighter and more manageable, as a template. To begin with we worked on the outside as a reference. Using blue tape as a horizontal guide we marked out the locations of the portlights (three on each side). In our case we could not do this symmetrically, due to the position of one of the interior bulkheads. You have to make sure that the portlights and the inner frames can be fitted without any interior interference (bulkheads, ceilings, wiring, air vents, etc.). By the time you have started to cut and drill it will be too late to avoid these problems.

➤ Without removing the protective tape, and with the aid of a template for the portlight made of cardboard, mark out the definitive dimensions of the hole to be cut. The straight section can be cut using the grinder as a kind of circular saw.

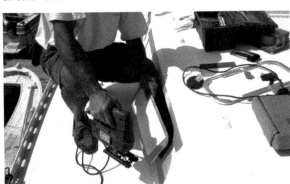

➤ We then used a jigsaw to cut round the corners. The protective tape which should still be in place will avoid the saw guides scratching gelcoat or paintwork.

➤ This next step required a certain degree of precision. We provisionally installed the portlight in order to mark out and cut, using a cutter, the overlap between the hole and the end of the frame. The rest of the protective tape is left in order to simplify the removal of the excess sealant.

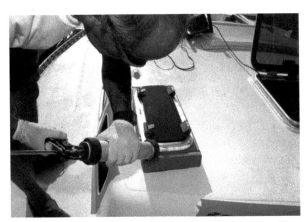

➤ Then we were all set to fit the portlight in place. First we squeezed a bead of polyurethane sealant around the whole of the frame.

➤ Next, first making sure not to install it upside down, we pushed the portlight slowly and carefully into place, pressing it down uniformly round the whole of its perimeter.

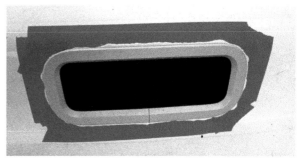

➤ The next day you can remove any protective strip and such sealant as has been squeezed out, using your Perspex chisel in the same way as you did for the deck hatches.

➤ Then, moving inside the boat, we fitted the inner frame/clamp ring. This will be your last chance to top up with any sealant that you think is needed; after this it will be too late. We then inserted the screws and tightened progressively, all around the inner frame. This part of Samba's cabin wall is curved, resulting in 2 mm of play between one end and the other of each portlight. In these cases you should never tighten the screws until the frame starts to give. This would seriously affect the watertightness of the portlight. You will have to rely on the sealant to fill the gaps.

➤ The new opening side portlights have substantially improved the look of the deck and, just as importantly, they will also vastly improve the circulation of air on the inside.

Replacing the Perspex panels

It is also quite easy to replace old Perspex windscreen panels, although a number of tricks can be employed to try to avoid the classic problem of your new panels breaking while you are working on them. The first is always to keep the Perspex sheets firmly clamped to avoid them vibrating. The second trick is to cut them using special aluminium saw blades; these are fine and have very small teeth. The third is to drill and cut the Perspex as close to right angles as possible. Finally, never ever use excess force while cutting or drilling Perspex: you have to let the machine work at it's own pace. Don't forget that the drill holes will also have to be countersunk on both sides. This is necessary to improve sealant watertightness.

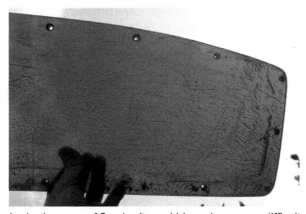

➤ In the case of Samba it would have been very difficult to replace these front panels with portlights due to the pronounced curve of the cabin. The old Perspex panels had become scratched and weathered over the years and the combination of sun and salt had left them virtually opaque. In short, a change was due. After about six or seven years Perspex inevitably becomes spoiled when exposed to sun, sea and storm. Perhaps the best way to extend the life of these panels is to protect them with a canvas.

➤ The best templates for the new panels are the old ones. Perspex with a smoked-glass finish is easy enough to find in just about any do-it-yourself store. Try to make sure that it is cast Perspex, rather than extruded, due to the superior quality; also reflected in the price of course. For tools you will need a jigsaw, a drill and the circular sander to smooth off the edges. When you have marked the outlines and holes all you need to do is get to work with the relevant power tool.

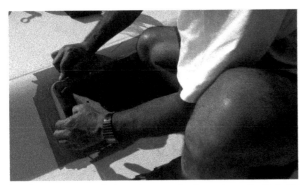

➤ Once new panels have been cut, all you have to do is install them. The first thing to do is protect the Perspex and the fibreglass, both inside and outside, with tape. This will avoid them being stained by the sealant. This phase is painstaking and rather tiresome but, in the long term, it is much easier than cleaning up the part of the sealant that squeezes out between the cabin and the panels. If there is a pronounced curve to the panels, as was our case where the curve of the front of the cabin is substantial, the best thing to do is heat them gently, or leave them in the sun for a few hours. That will soften them up, avoiding the risk of breakage when bending them to the required shape.

➤ The next step was to bolt them into place, with their washers and nuts, initially leaving them sufficiently loose so that the panel could be separated using shims (separators) or any handy screwdrivers. Another way of going about this, as we explained in the case of the hatches, consists of first applying the sealant to the deck and then, provisionally, lining the Perspex panel up using small screwdrivers, one at each end, as guides.

➤ We then applied a bead of sealant around the edges. This can be done working from either inside or outside the boat, whichever is more convenient. Remember that these panels require specially formulated sealants in order to resist the sun's ultraviolet rays without drying out or cracking: two problems that, in the mid term, are responsible for most water penetration.

➤ When we removed the screwdrivers, which had been keeping the two surfaces apart, the sealant occupied the whole outline, even getting squeezed out beyond the edges. The protective tape that we put on at the beginning simplified the eventual elimination of this excess.

➤ At this point we were ready to tighten the bolts. To optimise the properties of the sealant this should be done progressively and consistently all around the panel, without overdoing it. If the Perspex comes into direct contact with the underlying surface, apart from the risk of it breaking, this means that the layer of sealant has been squeezed out. By trying to ensure maximum watertightness you will in fact have achieved the opposite effect and, in no time at all, water will start dripping in. Ideally you should maintain a 1 or 2 mm gap for the sealant. Some professionals recommend an initial tightening, leaving it overnight and then tightening the bolts again the next day, when the sealant has set.

➤ Samba's new Perspex front panels now do justice to the rest of her deck hardware.

Chapter 22

Interior carpentry (part 1)

The most frequent image that comes to everyone's mind, whenever talk turns to the refitting of boats, is of carpentry work. Now why would that be?

Design and planning

For the redesign of *Samba*'s interior we reverted to her original layout, taking advantage of the existing structural elements and original furnishings wherever we could. The idea, suggested by some, of ripping out all of her innards, bulkheads and all, and starting from scratch seemed to me like madness. My idea was to adapt my requirements to the interesting possibilities that the existing interior layout had to offer. In the end our only 'invention' was the heads, previously almost non-existent, while for the rest all we did was improve the original furnishings.

If any readers are thinking of taking this chapter as a source of inspiration, I must warn you that today's production boats are not really given to modifications in terms of their furnishings. These days this kind of carpentry is highly industrialised, based on prefabricated units that slot into structural counter-moulds, whether in the hull or cabin ceiling, where the electrical wiring and water pipes are also hidden away. Introducing adaptations in these circumstances requires not only a lot of imagination, you also have to know exactly what you are getting into.

On the other hand, generally speaking, most production boats built before the early 1990s, offer a wider range of opportunities for refit. The carpentry is more classical, in the sense that it is based on wooden parts held together by screws and forming complete furniture units. This was the case with *Samba*, and many other boats of her generation.

Ships' carpentry covers a wide range of possibilities, working methods and materials. If we look at leisure sailing, from the old art and crafts approach of ships' carpenters in the late 19th and early 20th century to the present day mass production industrial systems, we can appreciate that modern production boats have achieved a truly polished look, with economies of time and money having been adopted in every aspect. Today there is an enormous range of technical resources and possibilities to choose from.

In our case we simply couldn't afford the luxury of cabinet makers, solid tropical hardwoods or complex handcrafted fittings. Neither was it possible, given the sophistication of their assembly, to fall back on the industrial systems employed on mass produced production yachts.

Our refit was simply based on plywood panels assembled onto a structure of wooden battens and screws. This is a relatively simple kind of carpentry yet one that can still provide a more than adequate finish. This system was ubiquitous in the seventies and eighties, and can still be seen in a few of present-day production boats. Marine plywood is a relatively cheap material, and well adapted to seafaring uses. It is light,

easy to work with, resistant to bending and loads, and ages well. You can buy it in a number of different quality wood veneer finishes (we decided on teak) which look pretty good when installed.

We have divided up the main carpentry work into three chapters. The first of these deals with cutting and fitting the panels, shelves and battens required to put together the basic structure of the furnishing units. In this first chapter we will also be looking at how to cut out the openings onto which the future doors will be fitted. In the second of these chapters we will be concentrating on mouldings, hinges, catches and varnishing, while the third will deal with the final assembly and fitting of all the varnished parts of our different furniture units. Finally, we will also be coming back to carpentry in the last chapters to look at all the final details and finishes that had to be left until the rest of the work had been done.

From the point of view of aesthetics, it is only at the end of the process, when you see the furnishings in all their glory, varnished and assembled, that you realise how good it looks. However, this present chapter, in which we designed, cut out and adjusted all of the main parts, was technically the most complex and also required the most time.

Rather than try to show you the construction modus operandi for each of the new units we thought it would be more illustrative to concentrate on the methodology. You will be able to see the final result soon enough; what we need to look at now is the process involved in this first, and important, stage.

Step by step

➤ We started to install the new furnishings by fitting and screwing into place the battens that would serve as the base of the panels. Before inserting the first screw, however, and this is the kind of work that cannot be shown in photographs, we spent a whole day with the carpenter discussing the pros and cons of every possible solution. You have to give this preparatory stage your full attention because this is where you will be making the main decisions. A carpenter will anticipate all those matters in which he is a specialist, such as the size of doors and the direction in which they will open. But you will also have to

allow for the taps in the heads, the light fittings, electrical sockets on bulkheads, the toilet, space for wiring and water pipes. You can deal with some of these aspects as you go along but thorough initial planning will save a lot of time in the long run.

➤ *Coming back to the battens, these must always be precisely fitted in terms of alignment and the height and possible inclination of each one, bearing in mind that, on a boat, no two surfaces are ever straight or parallel, even when they are supposed to be! The carpenter used Phillps screws to fix the battens in place, mainly because they are the ones that work best with power tools, vital for this kind of work.*

➤ For those battens that have to support greater loads, such as the work surface in the heads, rather than increasing the number or size of the screws we anticipated reinforcing the joints with polyurethane adhesive, a modern, practical and efficient solution.

➤ The tools required for interior carpentry work are basically a jigsaw, electric or hand planes and an electric drill/screwdriver. To save time it is a good idea to have two, or even three, electric drills set up, this means that you will not have to spend all day changing the drill bits. We also installed a fixed circular saw for cutting the biggest pieces, making minor adjustments in situ later, using the jigsaw. In order to simplify and optimise your supplies we recommend maximum unification of components. We limited ourselves to two types of battens, two thicknesses of plywood sheeting and two screw sizes.

➤ With the battens in place, the carpenter then went round making up the templates for marking out the plywood panels. He used MDF board for this, a pretty cheap solution, somewhere between wood and cardboard. This material is easy to work and allows for the fit to be refined. A good cost-saving tip is to reuse the larger templates to make smaller ones.

➤ The smaller templates were finished off inside the boat, which had been equipped as a workshop for the duration. With this kind of work you have to foresee this possibility as otherwise you will waste far too much time going backwards and forwards for each minor adjustment.

➤ When the templates had all been cut to size, the time came cut out the plywood panels. For Samba we used mainly 12 mm teak-finish marine plywood, also 15 mm for some load-bearing parts. On the templates you have to indicate which side will be the outside of the furnishing, fundamental for parts that are almost symmetrical, and also make sure the grain is straight, which is not always 90° to one of the sides. This is the kind of detail that you should always bear in mind.

➤ We then fitted the section of plywood in place and, as required, marked out and made the necessary final adjustments until it fitted correctly. The time had now come to move on to the next step which was to fit the shelves and door openings.

➤ In the head, the height and position of the toilet bowl was so crucial that it affected the design of all the cupboards and lockers on the starboard side. Once the position of the toilet bowl had been defined, including sufficient space for the seat to fully open and stay upright, we could start work on the rest of the furnishings.

➤ Little by little the battens start to mark out the outline and depths of the cupboards on the starboard side of the head.

➤ The false bulkheads that divide up the cupboards had to be adapted to the shape of the hull. There are various systems for defining their shape. In the photo you can see the carpenter using the batten and a tape measure with a spirit level attached, to mark out the distances every 10 centimetres on the vertical batten. This gives you a record of the successive distances from the batten to the hull. If you then mark out these distances on a template you will end up with a piece that adopts the precise curve of the hull. This system is both quick and reliable.

➤ The starboard side of the head furnishing had now started to come together. Here you can see the position of the toilet bowl and the adjacent bench, which will hide the taps, as well as the two back cupboards, one of which is going to be for waterproof clothing. We had also anticipated where the toilet filling and emptying pipes were going to run, making sure that they would be both accessible and hidden from view inside the cupboards.

➤ In the interests of space and functionality we decided against using a single piece for the back panel of the cupboards on the starboard side of the head. By facing off two pieces of plywood at an angle we managed to achieve good interior circulation, easy opening of the doors and to take best advantage of the available space, whilst respecting the overall aesthetics of the assembly as a whole.

➤ On the port side of the heads, the work top, where we would fit the washbasin, had already been installed and, in the photo, you can see the carpenter using battens to define the outline of the front panel.

➤ After fitting the central false bulkhead we then had to make up the shelves. This process was similar to that used for the previous steps. First we attached the battens at the required height, then we made up templates using MDF board and finally we marked out and cut the parts from the plywood sheets. For shelves that are never going to be much in view you could also use the off-cuts left over from other jobs, as the direction of the grain or any slight imperfections will be of little importance. You could also use marine plywood without a veneer, which is far cheaper than with a teak finish.

➤ Having fitted the parts of the new furnishings, we then removed them from the boat in order to cut out the door openings, fit the moulds, hinges and catches and varnish them. Whenever possible, it is better to varnish wood on dry land. Apart from avoiding smells, drips and possible varnish stains, this is the only way you can give a good even coat to all the nooks and crannies, including the edges and interiors.

Cutting the door openings

➤ The cupboard and locker assembly system that we decided to adopt takes advantage of the pieces cut out of the front panels to make the doors, with the subsequent saving in wood. To define the outline of each door we first marked it out in relation to the side walls, ceilings and false bulkheads using tape. Then, with the panel laid on a flat surface, we precisely and symmetrically marked the outline of each door. The carpenter did this using a set square, pencil and tape measure.

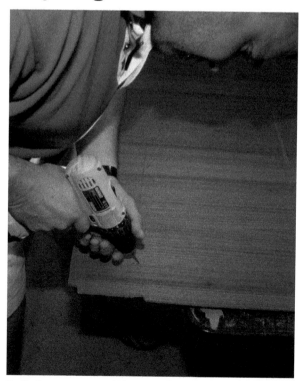

➤ Before you can cut these outlines out of the panel with a jigsaw you first have to open up a starting point that is wide enough for the blade of the jigsaw to enter. We managed this by drilling various holes, using a fine bit (2 or 3 mm), which we then converted into a slot by forcing the drill sideways.

➤ These small 'L' shaped slits at the diagonally opposite corners of each door allow the rest of the cut to be made using a jigsaw.

➤ When making these cuts, full concentration is essential, as it is very easy to drift off and lose the pencil line. If you are not careful the whole piece will be spoiled. However, any minor imperfections, of the kind that are unavoidable whenever you work with a jigsaw, will be covered over later by the frames.

➤ When we had cut out all the doors we marked each one with a label, indicating where they belonged and which way they opened. This step is far from redundant. Almost without realising it, we had made twenty door openings and doors for the new furnishings. Trusting your memory is far more risky and time consuming than simply labelling each door.

Creating a cupboard

➤ The four berths that originally formed part of Samba's interior layout ended on the sides of the fore cabin. After cleaning up this area we spent some time with the carpenter chewing over the best way to refurbish the new space that we had freed up so that it could be used for stowage purposes. The first thing we did was install a new panel, supported by interior battens at the bottom on the starboard side, copying the dimensions of its opposite number, on the port side.

➤ But in order to take advantage of the sides of the hull as a cupboard the top shelf would necessarily block off access to the lower one.

➤ As the spaces on one side and the other were quite deep we decided to make a double level of shelves.

➤ To solve this problem we made two openings on each side in the lower panels. (NB. The white colour of the panels that suddenly appears in this photo is due to the fact that the first phase of this carpentry work alternated with the interior painting.)

➤ *At the top of each side, the carpenter adjusted the front panels of the cupboards. The next step was to cut out the doors.*

➤ *Inside the cupboard, separated by a false bulkhead, you can make out the areas anticipated for hanging clothes and for shelves.*

Chapter 23

Deck hardware (part 2)

Little by little we had started to install the deck fittings. In this chapter we will be taking a look at how to install the new winches, rope clutches and deck organisers on the cabin roof, as well as the track for the genoa car.

Winches, deck organisers and rope clutches

An important aspect when considering changes to any of the deck hardware is to define the force that each new part will have to bear and, as a result, to choose the most appropriate items for the equipment. Manufacturers' catalogues and websites always include pages that will give you details of these forces, depending on your boat's footage or sail area. Shipyards, in order to shave their costs as much as possible, will often install parts with specifications that are lower than actually required, or will even underestimate the number of fittings needed. This is a problem that

new owners only become aware of once they have put to sea and come face to face with sprightly wind conditions, at which point they may discover that they need four arms to haul in a sheet or that blocks and pulleys start to creak alarmingly when subject to real stress.

In the case of *Samba* we opted for the comfort and security provided by slightly over-specified fittings, at the top end of the manufacturer's recommendations. Refitting a single boat, and determined to improve her rigging, the extra cost in terms of larger winches and blocks actually adds up to very little in comparison to the time spent installing them, whatever size they may be.

Basic advice

Installing deck hardware is, in technical terms, easily within the reach of the average enthusiast. Before taking on this work, however, there are a number of things that you should bear in mind. First of all, when drilling holes, always do so at right angles to the surface, if you do not, with the holes at an angle, the bolts will suffer by not working in the direction of their shafts, while washers will become twisted and might damage the structure of the deck. To avoid this problem, in hardware shops and do-it-yourself stores they sell plastic pieces, which serve as small guides, with holes at 90°, with the most usual drill bit sizes.

Another classic problem when drilling holes in decks are accidental knocks and scrapes caused by the drill's jaws or chuck. Whether drilling the deck directly or using part of the fittings as a template you should always take great care not to apply too much pressure, in particular at the point when the bit is finally about to break through to the other side.

Another frequent mistake among amateurs is to drill all of the holes in one go, having first marked them all out on the deck, one after the other. By doing it this way you will drastically increase the risk of some of your holes being slightly misaligned. It is far better to drill the holes one by one, fitting each bolt to check that everything is in line, before going on to the next hole.

It is also important to use sharp, high quality drill bits because, if you do not, there is a danger that the interior polyester will crumble and break, instead of leaving a clean hole. A piece of wood pressed tight against the underside of the deck while drilling may help this problem. Another indispensable accessory is the countersink, which will open up a small space so that the sealant can function as a kind of O-ring for each hole. Never use a large diameter bit as a countersink because it is almost impossible to control and you could easily end up drilling a large bore hole all the way through the fibreglass, causing resistance problems and/or areas for potential leaks.

Washers come in various sizes for each diameter of bolt (usually three). For deck hardware, particularly items that are subject to significant forces, it is imperative to use the largest size.

Do not over-tighten the nuts. Generally speaking, the best advice is to do up the different bolts progressively and to stop as soon as you can feel them start to bite into the fibreglass.

Speaking of bolts, always use high quality stainless steel; while it may seem trivial, before lining the item up and sealing it for installation, always check that the length, thickness and diameter of each bolt is correct, that you have the right nuts and washers and that there are enough of them.

Water tightness is another crucial aspect of all the elements that have to be installed on the deck. In this sense your best friend will be one-part polyurethane sealant. You should never forget that water only, and always, seeks its lowest level, never the opposite. It serves no purpose to fill the holes with sealant, as the bolts will simply drag the sealant down inside the boat, compromising its effectiveness and needlessly making a mess of hands and spanners or wrenches. It is better, depending on the instance, to use a little sealant at the base of the items, around the holes, or just under the heads of the bolts. In terms of the inevitable need to clean up traces of sealant, as some of it always gets squeezed out, the best thing is to remove the worst of it with a screwdriver or scraper and then clean off the rest with alcohol or turps; do not use acetone, which will damage both plastics and paintwork. When working with sealant it is more practical to use disposable paper for cleaning up rather than rags.

Finally one last piece of advice that, despite a thousand repetitions, remains one of the main causes of deck hardware installation disasters. Before installing the item, read the instruction manual and, before making a hole in your deck, make sure that the item is located precisely where it needs to be and that the holes on the underside are accessible in terms of fitting and doing up the nuts. Any mistakes of this kind are truly embarrassing.

➤ *Installing the different deck hardware components is a job that is easily within the reach of your average enthusiast. There are no major technical mysteries and no need for sophisticated tools or equipment.*

Step by step

Installing a winch

➤ When you have lined up the winch, mark the position of the holes using a pencil. Before doing this you should have already made sure that rope clutches and deck organisers are correctly lined up and, also important, that there are no obstacles to the crank turning. Do not forget that, before installing any deck hardware that will be subject to force, you will have to reinforce the deck. We did this by laminating sheets of plywood to the underside of the deck.

➤ It is important that you make all the holes at a perfect right angle to the surface; if not the bolts will not work correctly. Although all the holes are marked they should only be drilled out one by one, placing the winch in position and fitting the bolts after drilling each hole to make sure that it has not become misaligned.

➤ In order for the sealant to do what it is supposed to, it must be applied around the holes. If you put it actually in the hole it will be pushed down by the bolt, reducing its effectiveness and making a mess of hands and tools. The sealant will work much better if the holes are countersunk.

➤ Once you have the holes in place, surrounded by sealant, you can fit the bolts. These must first be inserted using a screwdriver.

➤ Tightening on the inside must be firm but not excessive. A good reference is the first sign of give in the fibreglass. It is also important to tighten the bolts bit by bit and alternatively, rather than individually and all the way.

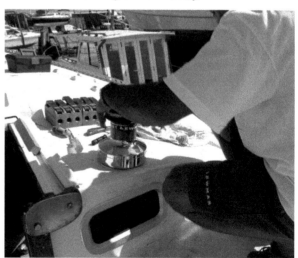

➤ The last stage of winch installation is to reassemble the drum. On Samba's cabin roof we installed two winches, one for the mast rigging and the other for the mainsail traveller.

Installing the rope clutches

➤ As with the winches, the first thing to do is mark out the position of the holes for each set of rope clutches using a pencil. On Samba we fitted four triple rope clutches, with maximum holding power of 1550 kg per rope. Once you have marked the holes drill them out one by one, fitting the bolts each time to make sure everything is still correctly aligned. In the photo on the left you can see the tube of the aspirator, held down in the left hand, to eliminate the dust as it is produced, simplifying visibility.

➤ You then fit all the bolts together, making sure the four rope clutches are correctly seated on the flat surface of the deck, and then progressively, bit by bit, do up the nuts. In this case due to the location you only need one person to hold the heads in place and tighten the nuts. Where this is not the case you will need two people to do the tightening.

➤ Once you have made and countersunk the holes run a final check to make sure the rope clutches all fit together and the bolts can be inserted without any problems. The next step is to apply the sealant around each hole, avoiding it being pushed through to the inside when you reinsert the bolts.

➤ Following the interior tightening, the four sets of rope clutches have been installed and are ready for use. You will inevitably have to clean up the excess sealant, some of which always ends up being squeezed down the holes or out from under the bases.

Installing the genoa cars

➤ The renewal of the deck hardware included new cars for the genoa traveller; replacing the old, short 'T' tracks that Samba used to be fitted with. We chose a model with bolts that were integrated into the profile, allowing us to take advantage of the original holes while the new tracks also hid the bolt heads from view. Our new tracks were 3 metres in length and fitted with bearings that were much smoother and more reliable than the friction bearings that had originally been fitted.

➤ As the new tracks were longer than the old ones we had to mark out the extra length of the traveller car, drill new holes and countersink them. Because we used a track with sliding bolts the precise distance between the holes was not of crucial importance.

➤ The next step, as always, was to apply sealant around the holes. This sealant would later be squeezed between the deck and the washers fitted to each bolt above the deck. It is important to seal off any possible point where water might filter down into the cabin.

➤ This stage, the hardest part of fitting the track, is a lot easier if you can muster up a couple of extra pairs of hands: one to hold down each end of the track and the third to make sure each bolt is fitted with its corresponding washer and correctly inserted into its hole.

➤ Unlike the rest of the deck hardware the installation of these tracks allows for the nuts to be tightened from inside the cabin without the need for anyone to hold the heads firm up on deck (they are held in place by the track itself). To simplify this task a power tool can be used, set to minimum power, for the first stage, finally tightening the nuts up manually, one by one.

➤ With the track installed we then fitted the end stops. In our case, with the aft end of the track almost in the cockpit we chose an end stop that included a traveller adjustment. To fit this we first had to drill some holes in the track. And then thread the holes using an M5 tap before inserting the grub screws. These holes are not ready-made as the tracks usually have to be cut to size (they come in 1, 1.5, 2 and 3 m lengths) depending on the size of your boat.

➤ After checking the threaded holes we could finally attach the end stops and its sheave and stopper to adjust the traveller car.

Installing deck organisers for halyards

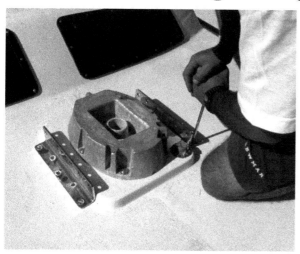

➤ The halyard outlets depend on these 'L' shaped flats for tying off the mast foot blocks. We had these made up for us but you can easily find them ready-made in any number of nautical hardware catalogues. Given the forces that they would have to resist we installed them using through bolts.

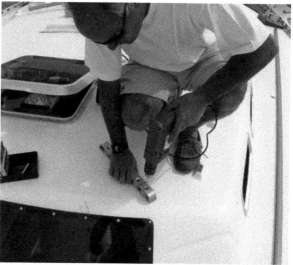

➤ Once you have decided on their position, mark out the holes using the deck organisers as a template and a small gauge bit, so as not to damage them.

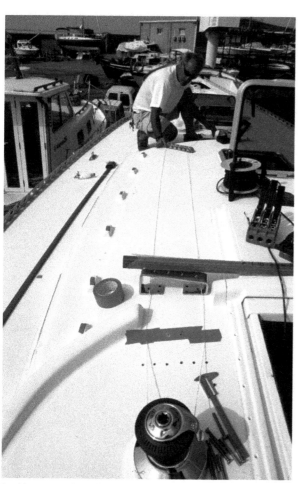

➤ You can visualise the correct angle for the halyard deck organisers using string to simulate their run. The deck organisers are in the best position when the lines run parallel over the deck.

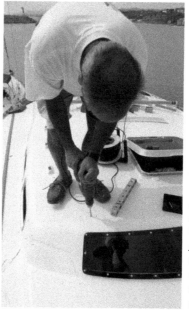

➤ Moving the deck organisers to one side, drill the final holes, making sure you always to drill at right angles to the surface and to avoid jamming the jaws and chuck into the deck when you finally break through to the other side.

➤ Now tape the area surrounding the deck organisers – this is in order to simplify the cleaning of the sealant that will be squeezed out. This item is now ready to be fitted and used. In the case of Samba we installed two deck organisers, each with six sheaves.

Refitting the original hardware

➤ From the original hardware we kept the genoa and spinnaker winches as well as the genoa traveller snatch blocks. While the winches worked perfectly, the wooden bases, once they had been cleaned up, were in a sorry state and in need of serious attention.

➤ One of the teak bases came apart while we were removing it but, with barely an hour's work and using the original as a template, the carpenter had made a perfect copy.

➤ After scouring the rest of the bases and renovating them with three good coats of varnish, we re-installed them in their original locations. These wooden bases, which these days have sadly fallen out of use, give Samba's cockpit a real touch of class.

➤ Installing the genoa traveller relay blocks turned out to be twice the work as there were two separate levels, both of which had to be sealed. First the assembly was fitted into place, loosely held down by two of its bolts, and then, in order to avoid staining, adhesive tape was attached to both the deck and the part of the wooden base that remains visible.

➤ The next step was to apply the sealant to the deck and fit the wooden base in place, followed by applying sealant to the 'second level'. On the wooden bases, as well as protecting the holes for the bolts, we also sealed the top and bottom edges of the wooden base. This will avoid water penetrating and getting trapped under either the base or the block and damaging them in the mid term.

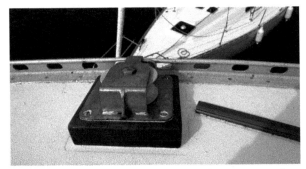

➤ Once the bolts had been made up, the excess sealant (after it had set) was scraped off using a well sharpened Perspex scraper and the tape was peeled off. The genoa traveller snatch block was ready for a few more seasons of faithful service.

Chapter 24

Interior carpentry (part 2)

In carpentry the difference between work done by an amateur and a professional can usually be seen in terms of the details: how well the doors fit, the neatness of the mouldings or the skill with which the varnish has been applied.

The value of experience

We finished the first chapter on *Samba*'s interior carpentry with most of the parts cut out and adjusted. Before finally installing the new furnishings we still had to attach the mouldings, hinges and catches to the doors and do the varnishing work. Varnishing is best done on dry land for reasons of convenience.

Apart from the technical questions, the first thing that has to be done by any amateur who wants to make up finishes in teak (or any other wood) on his boat is get hold of the corresponding mouldings and strips. Some chandleries have catalogues that offer a range of pre-cut teak angle pieces, moulds and baseboards in a variety of different sizes. Wholesalers and industrial carpentries also often have stocks of this wood and you can have the strips, planks and sections cut to size to suit the needs of a do-it-yourselfer.

Calculating your requirements and then checking out the prices is the first stage. Once more it is advisable to make up one single order for everything. This means you can negotiate a better price or, even more importantly with a single order, you can demand that the hue and grain of the wood be as near uniform as possible, which will result in an aesthetic uniformity of the finish.

With regard to the catches and hinges for the doors and lockers, there is an exhaustive choice. For *Samba* we chose brass trigger-catches, a system that is both classical and reliable, and which you will find installed in many production boats. We also decided on brass for the hinges because brass is a metal that stands up well to the marine environment and, over the years, oxidises to a dignified green hue.

Step by step

Mouldings, catches and hinges

➤ *The work of attaching the mouldings and fitting hinges and latches is similar for all doors. First of all you have to fit the two hinges to the panel/frame, making sure that both are recessed to the same depth; if not, the door is never going to open or shut correctly. You then have to fit the door to the other wing of the hinges, taking the same precautions as before in terms of symmetry. Also make sure that the door is not going to protrude and that it will close flush with the frame.*

➤ *After measuring the length of the moulding, always taking the longest side, cut both ends to a 45° angle. In hardware and do-it-yourself stores they sell guides that will allow you to correctly cut angles using a handsaw. In our case we relied on the precision of a circular saw.*

➤ *The first moulding to be fitted is the one on the hinge side of the door. In the photo you can see that there is a gap between the door and the panel, the mouldings will serve to disguise this gap. The gaps are due to the width of the cut (remember that the doors were cut directly out of the panels) and the recesses made to house the hinges. If you rebate the door by the thickness of the hinges this gap will later be hidden by the finishing mouldings, a small trick that is par for the course in large production shipyards and which avoids the need to slot the hinges one by one into each frame or use flat and delicate mouldings. In the photo we can see how the carpenter is drawing the only rebate required, the one that allows the moulding to slide over the hinges and cover over the gaps between frame and door. When fitting the hinges into the mouldings the best thing is to use a fine grade wood-file, cutter or small chisel.*

➤ *The carpenter used a two-part rapid-setting adhesive (the second part is a spray) for attaching the mouldings. Each moulding was then fixed with a pair of screws, inserted from the inside surface of the door, to avoid the veneer of the panels being damaged by accidental knocks.*

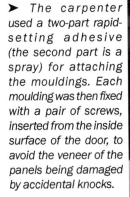

➤ The second moulding fitted to each door is the one opposite the hinges. It should be exactly the same length as the first and, when attached, you must make sure that the two are perfectly parallel. This is the only way to ensure that your frame will be rectangular.

➤ When the second moulding has been attached, the last two will have to be exactly the same (it is best to check this) and can be cut at the same time. These are then fitted, using the two-part rapid-setting adhesive (the same in all cases) and, when each frame is finished, inserting rear-mounted screws for security.

➤ Here the door is almost finished. As you can see, the mouldings, apart from improving the look of the door, hide the gaps between door and panel, and serve as stops when closing it. This means that you do not have to fit strips around the inside edges of the opening in the panel.

➤ Fitting the rings to line the holes is as easy as making the hole with a hole saw. The most important thing to take into account is that it must allow for the trigger-catch to be easily operated. Where you have two or more doors on the same panel you have to try to ensure the visual symmetry of the different holes and rings. There is no secret to fitting the rings, just carefully apply a little glue around the perimeter of the hole.

➤ Finally, as a finishing touch for the doors, gently round off the corners of the mouldings using a circular sander.

➤ When installing trigger-catches the carpenter will glue on wooden blocks (later inserting screws from the rear). The 'L' shaped stop of these catches has different adjustment sizes, to make sure that the doors close with precision.

➤ We finished the first door in a little over half an hour. Only 19 to go!

➤ When we had finished the carpentry work we took all of the panels and doors to be varnished, having first removed the catches and hinges. This is far simpler than masking them with paper to avoid varnish stains.

Sanding and varnishing

➤ We took advantage of using a number of the original baseboards, particularly those in teak or other quality woods. The old varnish finish on these had to be thoroughly sanded down to try to match the overall tone of the wood finishes. Over time varnished woods tend to darken, in the case of Samba her old teak baseboards looked more like ebony.

➤ Before varnishing the different parts they were all given a light sanding and were then blown-down to get rid of all the sawdust. Preparation and cleaning of the wood has to be meticulous. Varnish, rather than covering up any unnoticed finishing defects or specks of dust, tends to highlight them. We later applied a couple of coats of polyurethane primer, sanding the wood down again between each coat to improve adhesion and eliminate any small imperfections of the kind that always become evident whenever you varnish new wood. We diluted the first coat before application so that it would penetrate the wood.

➤ For the final finish we gave the wood two coats of two-part satin-finish polyurethane varnish, again sanded down between each coat. The photo perfectly illustrates the advantages of not doing this work on board the boat. Stripped down and with plenty of space you will have adequate access to both sides of each item. You should only consider doing the varnishing on board when stripping down the part or parts in question is not feasible.

➤ Once everything has been varnished, with all the items on the drying shelves, the carpenter will be ready to install them again in their respective locations on board.

Chapter 25

Interior carpentry (part 3)

Watching the new furnishings taking shape is a real delight. With the main structures assembled, the accommodation starts to take on the air of a true cruising yacht.

Setting the scene

Assembling the furnishings is undoubtedly the most impressive part of a carpenter's work yet, while it requires a certain skill, it is much faster and, comparatively, far easier than all the preparatory work that has been put in up to that point. The planning, cutting and adjusting of the new furnishings took us almost three weeks to do, while subsequently fitting the mouldings and doing the varnishing took us almost two weeks more. However, the final assembly shown in this chapter represented barely four days' work, including adapting and fitting a number of newly made mouldings and skirting boards.

Refitting a boat requires a lot of planning. The shipyards that build yachts have production assembly lines, where the hulls are moved round and the different specialists all do their bit (laminators, carpenters, painters, electricians, plumbers, etc.), everything according to a pre-established order so that each boat is put together in the easiest and most effective way in terms of the different specialists or work teams.

In the case of *Samba* we had to try a similar system, to ensure that each group of specialists could move in and do their bit in the right order and at the right time. The sequential structure of this book may give the impression that the work went ahead in fits and starts (which is only true up to a point) but the alternating of specialists did respond to a certain pre-established logic. Each stage of the work was dependent on what had been done before and determind what would be done next. Progressing through this apparent disorder, contrary to what may seem to be the case, resulted in a significant saving of time. Above all you must never forget that the most expensive part of any refit is the cost of skilled labour.

Getting back to the carpentry, although the carpenter may appear to have finished his contribution to *Samba*'s refit with the work shown in this chapter, this is, in fact, far from the case. Later on you will see him again, renovating the side panels, doing the cabin floors and working on the final finishes. The carpenter will be back with us again: he cannot finish his work until everyone else has finished theirs.

Step by step

➤ One of the first steps is to firmly anchor the battens that form the supporting structure for panels and shelving. Greater rigidity is achieved by a combination of screws and polyurethane adhesive. This type of adhesive ensures a very strong joint while preventing humidity seeping in behind and under battens. Polyurethane also offers a certain flexibility, avoiding the classic 'creaking' of furnishings whenever the boat twists imperceptibly as a result of the combined action of sea and wind. In the long run these flexions will end up loosening screws or compromising the adhesion of more rigid adhesives and are the origin of problems involving badly fitting doors and shelves, or loose joints and hinges.

➤ Battens that are directly attached to the hull (others are attached to bulkheads) must be completely reliant on polyurethane adhesive, which even gets used to fill the gaps left by the curved shape of the hull. Fitting battens, panels and bulwarks is a widely used system in the building of production boats that is solid, quick and easy.

➤ To fix the larger shelves, like this one at the bottom of the forward cabin side cupboard, you should first apply a bead of sealant, particularly in the corners and the holes for the screws, before inserting the screws. This will ensure that the whole of the shelf is uniformly seated, which is important for large pieces like this.

➤ For the smaller shelves no screws are necessary, just two or three dabs of sealant.

➤ With a couple of thumps, these shelves fit into position in a solid yet slightly flexible way. Should it be necessary to remove them (although we hope this will not be the case) instead of having to loosen screws you just have to slide a cutter blade between the shelf and the supporting batten. It is almost impossible to separate joins made with polyurethane on the basis of brute strength alone.

➤ The front panels of the cupboards are assembled in a similar way. After fitting the screws in their holes (from the rear) the carpenter applies a bead of sealant, or a series of small dabs, around the whole of the batten to which the panel is attached.

➤ The panel is then fitted into place, making sure that it is correctly aligned both vertically and horizontally.

➤ The front panels are finally fixed in place with screws inserted from the rear of the batten, which avoids their heads being seen. You also have to look after the aesthetics!

➤ With the panels in place, the next step is to fit the doors. We removed all the hinges and catches to ensure a better varnish so the first thing we had to do was to put them back. As all the holes were made in the previous stage the work of assembling and fitting them did not take very long at all.

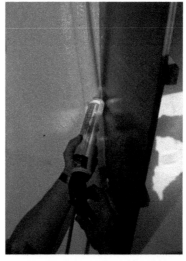

➤ Joining the false, vertical bulkheads to the hull is done directly, only using sealant (no screws). In this case we used the strongest available polyurethane adhesive, the same as is used by some shipyards to join decks to hulls.

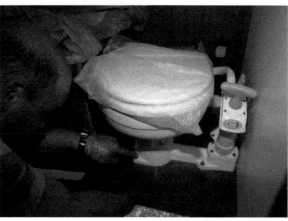

➤ The moment also arrived to finally install the toilet bowl. Four bolts and a bit of sealant around the base is sufficient for this purpose.

➤ The installation of the furnishings (although the final finishing work will have to wait until later) was brought to an end with the fitting of the main decorative mouldings and skirting boards. We were unable to restore all of Samba's original mouldings, some of which had come apart as we were removing them, while others were in such bad condition that we had no choice but to replace them. However, this left us with the problem of getting the colour of the new pieces to match that of the old pieces.

➤ *The know-how and experience of the carpenter was essential when it came to copying the profile of the old skirting boards and mouldings. However, the problem with the colour was a more complex issue. The same type of wood may present a wide range of different shades, depending on age, country of origin or tree it was cut from. As we disliked the idea of using artificial wood dyes, we decided it would be better to leave the matching of the wood tones to the hand of time. What we did try to do, for aesthetic reasons, was try to match up the different shades depending on where they were being used.*

➤ *For the finer adjustment of the skirting boards and mouldings the carpenter set up a temporary workshop in the saloon. As the parts to cut were not too big it was far easier to bring the circular saw onboard than be constantly toing and froing to adjust every cut.*

A positive change

These last four photos give you an idea of what we achieved with *Samba*'s interiors. The combination of natural wood tones against the white background of the bulkheads was standard for yachts from the beginning of the last century. In our opinion this combination continues to offer an attractive and bright contrast that has added to the sense of space in *Samba*'s interior.

The photos also show the panel behind the sink, with the mirrors fitted to the doors. We would like to take this opportunity to remind DIY enthusiasts that mirrors have to be glued into place using adhesive specially designed for the purpose. If you do not do this, you run the risk that the backs of your mirrors will rapidly oxidise and blacken, a problem that many people mistakenly put down to humidity. The easiest thing to do, whenever possible, is to remove the door and take it to a glazier so that the mirrors can be cut to size and properly attached.

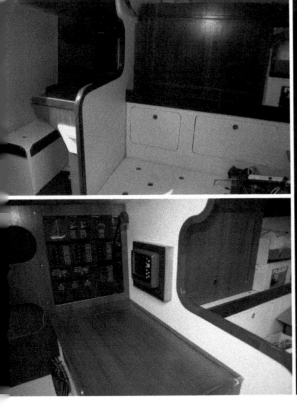

Chapter 26

Installing the electronics

Nowadays there are few sailors who are prepared to go to sea without electronic backup, and we are not among them. So for Samba we decided on a multifunction display system, an autopilot mounted below deck and a combined chart plotter/GPS/sounder.

Here's to modernity

Nautical electronics is advancing in leaps and bounds, hardly anything that we were going to install in *Samba* had even been invented when she was launched thirty odd years previously and, by the time this book comes into your hands, it is likely that some of the instruments that we installed will have become obsolete. With this in mind, rather than going into great detail about the specifications of these instruments – which will inevitably be rendered obsolete by the passage of time – we are going to occupy ourselves with the criteria that we followed for their selection – a much less changeable aspect.

Many shipyards offer new boats equipped with electronic packages, an option that will give you advantageous prices and simplified installations, the wiring is also neater and much better organised than it would be if you had to install it yourself. It was precisely this approach, simplicity of installation, that we wanted to apply to *Samba*, by coordinating the installation of the electronics and electrical wiring. In this way we were able to lay the wiring for sensors and repeaters along the same routes as the electrical wiring, hidden from view yet accessible under lockers and behind cupboards.

Compatibility is another important factor to be considered when deciding on your electronics. The more complex the installation (plotter, depth sounder, autopilot, multifunction display, radar, PC, TV, satellite, etc.), the more sophisticated the communication between the different apparatus has to be. If your intention is to gradually augment and extend your electronics then you should consider choosing a particular brand and sticking with it (simplicity of interconnection) or, at least, installing equipment where interconnectability is guaranteed.

In principle installing the electronics – above all the multifunction display – is a job within the reach of most amateurs. Yet, as the equipment gets more complicated, when it starts to enter into the sphere of information technology, you are probably going to need the services of an expert in the field.

When discussing electronics it does not make much sense to talk about good or bad instruments. Where some people only require a few basic functions at an affordable price, others demand the best and do not care how much it costs and yet others just want equipment that is easy to use. Each owner has his own requirements and must make his own choice.

Other factors, which we might term functional, also enter into play. The most important of which is *Samba*'s role as a leisure sailing boat. This being the case it appears to be logical that the multifunction display and the autopilot should be the keystones to which the other components were added, which was exactly the way we did it, until we had designed a complete electronics package. In the case of a fishing enthusiast, for example, it would be normal to start with graphic sounder and fish-finder, to satisfy the owner's particular requirements, and then build up the rest of the electronic equipment from there. Following this procedure will simplify the choice and avoid making compromises in terms of the equipment that is going to be of most use once you have got your boat back on the water.

Step by step

Installing the autopilot

➤ We started off the installation of our autopilot by making up this wedge-shaped wooden chock, which would serve as the base for the directional piston. To make this piece up we combined layers of solid wood and plywood, to ensure maximum hardness and a perfect grip, on the plywood, for the piston support screws. The exact dimensions and angles required for this chock were calculated by the electronics technicians, while the laminator actually made it up and installed it.

➤ To install the wooden chock/support we first glued it to the hull using fibreglass paste. Before doing this we sanded down and prepared the surface of the hull where it was to be installed, to make sure that the paste and subsequent laminates would adhere properly.

➤ You can see the wooden chock that will support the piston in the background and, in the foreground, there is another similar looking chock that will not be subjected to any force, as it only has to support the hydraulic motor in a horizontal position.

> It took a real expert to laminate up to ten solid layers of fibreglass. To a great extent, the autopilot's ability to unflaggingly perform its task over the years will depend on the rigidness of this chock. You must also take into account that the autopilot of a boat such as Samba (a 40-footer with a displacement of 9300 kg) is going to be subject to stresses in excess of a hundred kilos.

> Once the chocks had been laminated firmly in place we started work on the installation of the servo unit itself. Given the complexity of this work some may consider that it would have been much simpler just to install a cockpit pilot. Cockpit autopilots act directly on the wheel and are much cheaper and easier to install as their mechanism and controls are in full view. The problem was that an above-deck autopilot would be at the limit of its potential on a boat with Samba's displacement. All the experts agree that if you are going to invest in an autopilot you should aim high rather than low in terms of power, in this way guaranteeing that it will not decide to pack up on you when you need it most.

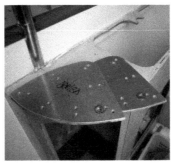

> At the same time as we were laminating the supports, the electronics technicians made up these aluminium pieces from cardboard templates which woud serve to structurally reinforce this sector of the rudder. As mentioned above, autopilots, particularly those mounted below deck, have to deal with enormous loads in order to control a boat in all possible conditions and, consequently, your whole steering system will have to be reinforced.

> Screwing the piston and the hydraulic motor into place on the chocks is a relatively simple task, compared with the work that has been done up to now. Fitting a below-decks autopilot often borders on the impossible. Most boats have a space set aside for electronics alongside the chart table, or in the wheelhouse, but fitting a below-decks autopilot is something that will rarely have been anticipated in production boats, where you will usually have to start from scratch, making use of whatever space can be found in the interior helm area.

> In this photo you can see the installation when it had been almost completed, although you cannot do the final calibrations until you have her back on the water and the central computer, the true brains behind the adjustments and automation of this instrument, has been installed. The autopilot 'black box' visible on the left of the photo, was screwed into place on a piece of marine plywood that had previously been glued to the hull using polyurethane adhesive. The motor-piston-computer assembly is connected up in a neat and orderly fashion, with all the elements grouped together and accessible for such servicing as may need to be done at a later date.

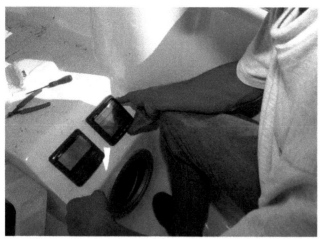

➤ The only part of the autopilot visible on deck is the repeater, alongside the steering control with a central dial or buttons. The LCD screen shows the main data for the course being followed or the waypoint, as well as an analog CDI showing the position of the rudder blade. Thanks to an interconnection multifunction display unit, data can also be seen on the data screen.

Installing multifunction display

➤ Samba's original electronic equipment, selected from among the best of what was available thirty years previously, was just begging to be replaced. Nothing was working as it should and, although we could have tried to repair it, the spare parts and labour costs would have set us back almost as much as replacing it with new equipment eventually did. Sentimentality apart, the specifications of modern equipment are far superior and far more reliable.

➤ A significant part of the electronics installation was fitted at the same time as the electrical wiring, with all of the repeater and sensor wiring being laid along the same ducts as the electrical wiring. This resulted in an altogether much neater and tidier installation.

➤ The best location for the electronic compass sensor is on the side of one of the forward cabin lockers. These sensors have to be installed as far as possible away (at least 1 metre) from any metallic mass or magnetic interference. In a sailing yacht the keel and engine are the main dangers to be avoided, although you must not forget televisions, speakers or even mobile phones, which are all capable of altering a compass reading by several degrees.

➤ We decided to locate the multifunction display 'black box' behind the switching panel, which is where all the repeater and aerial, autopilot, transducer and chart table display wiring is discreetly brought together.

➤ We installed the GPS aerial on the lower railing of the bow pulpit, an uncluttered yet relatively well protected area which allowed us to take advantage of the run of an old VHF cable that had previously occupied this same location.

➤ It was quite easy to install the interior connections of the four repeaters (pilot, wind and two multifunction displays) as access to these was possible directly from the aft locker or aft cabin. In some modern boats these connections entail wiring requiring all the skills of keyhole surgery.
➤ In the cockpit the four repeaters (two on each side)

are housed in the chamfered space at the aft end of the cockpit benches. When Samba was built, rudder consoles with instrument panels had yet to be invented.

➤ We completed our electronics installation with the plotter/GPS/sounder installed alongside the chart table. The particular aspect of this apparatus was its independent chart housing, which allowed for it to be assembled in places where there was no risk of water entering, accidental knocks or mishandling. In this way the delicate magnetic card connectors were protected inside the chart table itself.

➤ We built the GPS/plotter display into the cover of an old water-colour paint box, duly restored and attached to the bulkhead using a length of continuous (piano) hinge, salvaged from one of the scrapped panels, and with a top-mounted plastic pressure-catch. This 'invention' meant we did not have to make any holes in the bulkhead to build the instrument in and made access to the back highly practical. We also found it a far more elegant solution than the usual U-shaped metal brackets.

Chapter 27

Renewing upholstery and foam fillings

Beyond the issue of aesthetics, when it came to renewing the upholstery we encountered endless technical issues. When you add to this the vital importance of foam filling, you find you are actually dealing with a far more complex question than you first thought, yet one that is essential in terms of the quality of life on board.

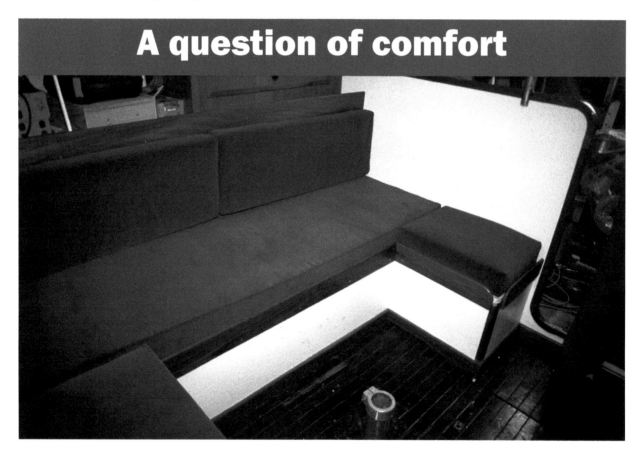

A question of comfort

Most of the improvements that end up getting made to boats are in response to practical necessities; parts tend to get broken and have to be replaced. Renovating the upholstery, on the other hand, is something that, generally speaking, is motivated by emotion, for the purpose of making life on board more comfortable and giving your boat that personal touch.

Nobody can deny the radical improvement that the renovation and updating of her upholstery will represent for the interior of any used boat. In terms of the comfort of life aboard this is an investment that is

more than justified. Upholstery and foam fillings do not last forever; they always end up getting dirty or spoiled or both and, little by little, become less comfortable and more unsightly. When the time comes to renew them many people are surprised by the multiple options available to them, and also by the cost of some of these options.

Fabric: A wealth of possibilities
The choice of upholstery fabric, along with the corresponding combinations of colours and patterns, is virtually infinite. But they will not all be sufficiently hard-

wearing or well adapted to the nautical environment. Once you have taken the decision to replace the upholstery and foam fillings of your interior furnishings it would be a shame to spoil it by using poor quality or inappropriate materials. The price of the materials is only going to represent a small part of the final bill, in which, yet again, the main part will be the cost of skilled labour. Each fabric has its advantages and disadvantages, and in the following paragraphs we will attempt to summarise these briefly.

Microfibre: Undoubtedly the best choice for boats; shipyards that use microfibre tend to boast about it. Microfibre is manufactured by a number of companies, although it is often referred to by the trade name of one of these – Alcantara. It has a smooth, velvety feel and is very resistant to wear and tear. It can be washed as often as required, at up to 30°C, and will not shrink or lose its colour. These fabrics are also fireproof, resistant to static electricity (do not attract dust or mites) and incorporate a layer of Teflon that makes them resistant to both water and staining. Following this list of virtues their only disadvantages that come readily to mind are that this fabric only exists in single colours, prints at most, and that the price is a faithful reflection of its qualities.

Acrylics: This is an interesting compromise between quality and price for nautical upholstery. Acrylic fabrics (Dacron is one of the best known) are easy to make up, age with a certain dignity and can be found in a great variety of single colours, patterns, checks, stripes, you name it... Acrylic upholstery can be machine washed, although for best results in cold water, with no need to take great care.

Cotton and viscose: While these are both great for home use they are not recommended for life at sea. When new their warm and natural feel is great, but the fibre absorbs humidity and mildew with far too much ease, and can quickly become a source of unpleasant smells. Cotton also shrinks – up to 5% in the wash – complicating the task of reinserting your foam fillers. You will find the same problems with viscose, or even worse; it is capable of shrinking up to 20%. The best advice is to avoid these two completely, particularly given that the cost is practically the same as acrylics.

Leather: Most unusual in production boats, leather is a very resistant material and much appreciated by boat owners who want to give an touch of class to their saloon. When used for nautical upholstery, leather must be given a damp-proof treatment, similar to that given to shoe leather. Wiping down with a sponge dipped in fresh water is the only care that it will subsequently need; avoid cleaning products that will only impregnate the surface of the leather and harden it. Before deciding on leather it is a good idea to check out the prices, which tend to be discouraging.

Synthetics: This group includes a wide range of different fabrics using composites such as Skai, vinyl, neoprene or PVC. These are often used for cushions in the cockpit area of a wide range of boats and also in the interiors of day-boats and sports or charter boats. In general synthetic fabrics are outstanding in terms of their resistance to wear and tear and ease of maintenance. However, you do have to make sure that they have been prepared for a marine environment and exposure to ultraviolet rays. You also have to choose the correct weight or thickness (neither too much nor too little). Depending on their composition, you can find them in all price ranges. The main disadvantage with their use, above all in summer, is that they do not usually breathe, resulting in the unpleasant sensation of your bare skin sticking to the cushions.

Foam fillings: Looking for the best compromise
The foam fillings used on board a boat are subject to different requirements from household mattresses. The first of these is, generally speaking, that on a boat there are no bed-bases. Bed-bases serve to cushion domestic mattresses, absorbing a third of the user's movements. On board a boat your foam mattress will rest directly on top of a sheet of plywood, which means that the thickness and hardness of the foam itself are the only factors affecting user comfort.

In order to guarantee your comfort, the foam must combine a number of qualities that may, on occasions, be contradictory. On the one hand it has to be hard enough to avoid you feeling the wooden base and yet, at the same time, it must be soft enough to ensure your comfort. Thirdly, and no less importantly, it must also be capable of dissipating your perspiration (up to half a litre per night) and body heat.

Mattresses used on board a boat are usually made of polyether or polyurethane foam. These foams are manufactured in large blocks which are then cut to any thickness and dimensions required.

With foam, one of the reference qualities is its density. A foam with a density of 30 – using 30 kg of material per m³ – will generally be quite soft and has less capacity for recovering from deformation than a foam with a density of 40 or 60. We say 'generally' because there is also the factor of compression hardness, which is defined by a different index. This means that, for the same density (capacity for recovery), you can find foams of different levels of hardnesses (comfort). As you may have figured, a cubic metre of foam will increase in price to the same extent as its density increases and hardness is reduced.

A good compromise for boat berths is probably using a foam with a thickness of 100/120 mm and a density of 30 to 40 kg/m³, depending on hardness. The greater the density, the lower the hardness can be. Things are more complicated in the saloon or the fore cabin, where it may be necessary to use denser foam to avoid a sinking sensation on the benches. A person weighing 80 kilos when sitting places three-quarters of that weight on the seat, only 6 to 9 kilos on the backrest (remember to use softer foam here), and the rest on his feet.

Latex: Investing in comfort

Latex is a derivative of a liquid extracted from a tropical tree. When converted into foam, in addition to its other qualities, it also has the capacity to adapt to your bodily temperature and offers the comfort of its impressive powers of adaptation and excellent elasticity. Latex mattresses are also anti-allergenic and anti-static. Disadvantages include weight (65 to $90\,kg/m^3$), price – three or four times more expensive than polyether foam, and sensitivity to ultraviolet rays. The nautical latex market is mainly for boats over a certain length, which may be at sea for several days or even weeks in a row, and where the owner is prepared to make a sizeable investment in comfort. Finally, there are mattresses that combine layers of polyether and latex and may even include natural fibres (in particular coconut). The advantages and disadvantages of these sandwiches are a reflection of those mentioned above for the different materials that make them up.

A few loose ends

✓ Whatever material you choose for the new upholstery you must use polyester thread, or similar, to sew the seams.

✓ Ensure the material and/or foam you choose is fire retardant and complies to current fire safety regulations.

✓ Any buttons or other decorative elements that you might want to add should not be made of metal (use plastic, wood, mother of pearl, etc.). Even stainless steel will leave rust marks on the fabric.

✓ Although you can use the old foam fillings as templates for the new ones you should always check these out on the boat first; they could easily have lost their shape over the years.

✓ Plastic zips or strips of Velcro are the recommended option for closing the covers. It is also a good idea to clean zips with fresh water at the end of each season, especially if you think they may have come into contact with seawater.

✓ Should it be the case that the owner, or a close and exceptionally well-intentioned member of his family, is not particularly handy with a sewing machine, it is far better to leave this work in the hands of a professional. As long as you have a clear idea of what fabrics, foams and accessories you are going to use, there is no need to go to a so-called nautical upholstere; this is work that any competent upholsterer can handle.

✓ The best way of getting air to circulate through the foam, and avoid condensation due to humidity and sweat, is to make holes in the locker lids, which will also prevent bad smells accumulating in the lockers.

✓ Leaving the mattresses and covers standing upright through the winter is another way of avoiding condensation and unpleasant smells.

Step by step

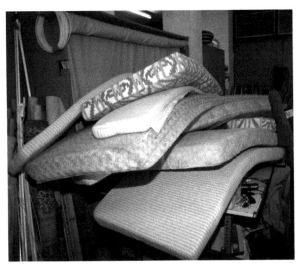

➤ Samba's *old upholstery and foam fillings, worn down and falling apart, desperately needed to be replaced. Renovating the upholstery is something that almost always gets done for reasons of emotion, through a desire to keep sailing the same boat for many years.*

➤ *Although old foam fillers can be used as templates for the new ones, you should always try them out first on the boat. Over time one or more of your old foam fillings may have lost their shape.*

➤ For the cushions that are in direct contact with the hull, as is usual with berths, apart from the possible deformation, you should also make sure that they are cut to right angles.

➤ Polyether and polyurethane foam fillers are manufactured in large blocks. They are colour coded, which means that you can visually identify the hardness of each one.

➤ A special fretsaw, of enormous dimensions, is used to cut the pieces from the blocks, as required by each upholsterer, although small adjustments may be required before they can actually be fitted.

➤ Latex has a capability to adapt to body temperature. This ability to adapt, along with excellent elasticity, ensures a high level of comfort. Latex mattresses are also anti-allergenic and anti-static.

➤ Generally speaking it is a good idea to take upholstery colour charts or sample books back to the boat so that you can see what your chosen colours look like in situ and with natural light. For Samba we decided on microfibre in a burgundy tone (for the saloon) and off-white (for the berths): two classic colours that combine well with the rest of Samba's interior decoration. However, in questions of taste there are no rules.

➤ The foam fillings used on the berths were 10 cm thick, with a hardness of 40, while for the backrests they were 7 cm thick with a hardness of 30. The upholsterer checked all the measurements before cutting them to size.

➤ *Cutting out and sewing little by little, the covers started to take shape. Whatever material chosen for the upholstery, the seams should always be sewn using polyester thread, or similar, never cotton thread.*

➤ *Buttons, zips or any other decorative element that you may decide to include must be synthetic and not metal as even stainless steel will leave rust marks on the fabric.*

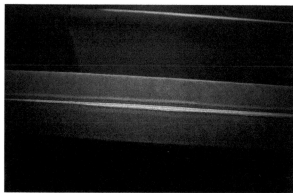

➤ *Leave the zips, for removing and replacing the foam filler, in the centre along the hidden side of each cover. This will allow you to turn any symmetrical cushions and mattress covers over. For greater comfort the backrest cushions have been covered with a synthetic mesh.*

A change for the better

The improved look that the new upholstery has given to *Samba*'s saloon and fore cabin is plain to see. After these photos had been taken we again removed *Samba*'s new upholstery; there was a lot of work still to do and it would have been a shame to get it dirty before we had even finished.

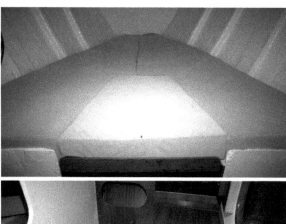

Chapter 28

Installing the refridgerator

With a bit of skill and without the need for any special tools everyone can install their own on-board refrigeration unit, although people usually tend to leave this job to a professional.

On-board cold

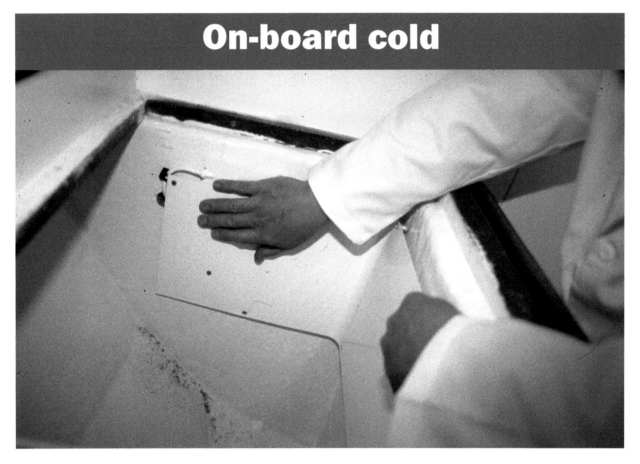

Few boats of 30 feet and over leave the shipyard these days without some kind of refrigeration unit, either as a standard or optional fitting. A quarter of a century ago this was not the case, and on-board refrigeration was a luxury reserved for the self-indulgent. The old refrigeration units were also an endless source of technical problems and spare parts were expensive and hard to find. Fortunately things have improved immensely and technological advances have managed to overcome the bad name of those old units.

Most boats have space set aside for the installation of a refrigeration unit. If this is not the case you will have to start by creating one: think in terms of complete 12 V cold boxes or even portable units. Cold boxes that have been designed to be installed on boats are similar in technical terms to domestic fridges. They are fitted with a compressor, an evaporator and a radiator, all of which are joined together by wiring and pipes. The main difference is that on a boat the cold box has already been mounted and all you have to do is install the refrigeration machinery.

The most usual on-board refrigeration units work by compressing gas (usually freon) until it becomes a liquid and then allowing it to evaporate inside the cold box, which produces cold as a result of the heat exchange. There are also Peltier-effect cold boxes, particularly for portable units. These cold boxes, which can produce heat as well as cold, are based on an effect discovered by the French physicist, who observed that electricity passing through two different materials would produce cold or heat, depending on the direction of the current. The Peltier system allows the temperature of the cold box to oscillate between 20/30° above or below the ambient temperature.

Other cold boxes work by absorption, a system based on the gasification of a liquid (usually water and ammonia) using heat produced by a flame, which in turn produces cold on returning to its liquid form. The main disadvantage with this system, which was in wide use years ago, is the worry of having a flame constantly burning on board. As a result this system has now gone out of fashion. It is the usual system in triple supply units (12 V, 220 V and gas).

In the case of Samba, as with most yachts of her size, we decided to go for a compressor system. The first thing we did was to measure the volume of the cold box, which turned out to be more or less 150 litres. The second step was to measure the size of the cold box walls, to find out where and how we could install an evaporator plate that would be adequate for the volume of our cold box, as well as a prior study of the best location for the compressor – somewhere as low, well ventilated and protected as possible.

Once we had this information we then looked for a unit that was perfectly adapted to our needs, especially with regard to the 'L' shaped evaporator plate that we wanted

to install, and with dimensions that were almost made to measure for our cold box. The installation, which was done by a refrigeration fitter, represented about three hours' work. A small investment for the great benefit of having an electrical cold box on board.

➤ Portable cold boxes are a good alternative for cruising yachts, whether as first choice on smaller boats or as an addition on larger ones.

➤ Front opening cold box/fridges are the most common option for motorboats and have also become quite popular for sailing boats. While they look just like small domestic fridges they are made out of materials that are resistant to a marine environment. On sailing boats (except for catamarans), the door may open and spill the contents out when heeling.

Refrigeration units: The basic elements

The heart of a cold box is its compressor, in charge of compressing and circulating the refrigerating liquid/gas. When compressed the gas liquefies and is delivered to the evaporator, inside the cold box, by means of a narrow copper pipe. Inside the cold box the gas liquefies again, exchanging heat with the hotter air it encounters, and is returned once more to the compressor to be liquefied again. It is just like water boiling in a pan, with the difference that freon turns to gas at -10°C rather than 100°C. The heat inside the cold box is the flame that makes it boil.

The evaporator, bolted to the inside of the cold box, is the most visible part of the refrigeration unit. These plates may be flat, 'L' shaped or in a closed box to allow for ice-cubes to be made. There are also eutectic models, which are sealed stainless steel boxes filled with a liquid that stores cold and releases it little by little throughout the day. There are few differences in terms of performance between eutectic evaporators and traditional ones. The former, for example, will require the compressor to work continuously for two hours during the morning, later releasing the cold throughout the day without further electricity consumption. Traditional evaporators will also make the compressor work for about two hours a day to achieve the same level of cold, although in this case they will do so intermittently throughout the day. The choice between one and the other will depend on each owner's individual needs.

The condenser is another important element in a refrigeration unit, serving to evacuate the heat of the gas as it travels between the evaporator and the compressor. The most usual types are small radiators with fins attached to the compressor, although there are also more complex systems that take advantage of the coldness of seawater to achieve a cooling effect. In these cases the condenser is shaped like a transducer and some models can be installed to take advantage of an existing through-hull, usually the drain of the kitchen sink.

These three main elements are complemented by the capillary tube and the thermostat. The first is a small diameter spiral-shaped copper tube, the purpose of which is to increase the pressure of the liquefied gas that is supplied by the compressor so that it will suddenly expand once inside the evaporator. The thermostat, as its name indicates, is a control for regulating the temperature of the cold box in which it is installed.

A few loose ends

✓ We recommend that, when choosing a refrigeration unit, you go for slightly more output than you actually need. For example, if your boat's cold box is 150 litres it is better to look for a unit that can handle up to 180 litres, but avoid exaggeration. A unit that gives you twice or three times the necessary output will not mean that your compressor has to work for less time and will not use less electricity. In the end more powerful motors need the same amount of energy as smaller ones to achieve the same temperatures and if your motor is too powerful you could run the risk of freezing the exterior evaporator pipes as the fins will not be able to evacuate all of the cold produced by the quantity of gas circulating.

✓ A portable cold box is a good alternative for cruising, whether as the first option in smaller boats or as a back-up in larger ones. You can find portable cold boxes that work with a compressor, gas or the Peltier system. When these cold boxes reach a certain size, the only portable thing about them is their name – they can be really heavy. In these cases we recommend models with wheels.

✓ A simple, cheap camping cold box, for keeping your drinks cool on a day out with the aid of ice cubes, is also a saving in terms of both amps and batteries, as it avoids constantly having to open and shut the door (loss of cold) of your electrical cold box.

✓ Over 80 per cent of the compressors on the market are manufactured by a company called Danfoss. The different makes of cold boxes then build their systems around this part, adding different peripheral accessories depending on the models (evaporators, condensers, electronic thermostat equipment, connectors, bodies, etc.).

✓ When deciding on the output of a refrigeration unit you have to define at what longitudes you will be sailing, and during what time of the year. It is not the same to cool a can of beer to 4°C in winter in Patagonia as it is to do the same thing in the Caribbean in August.

✓ Front opening rectangular cold box/fridges (you can find them with the refrigeration unit already built in) are an interesting option for a motor boat. While they look just like small domestic fridges they are built using materials that are resistant to a marine environment. For sailing boats (except catamarans) the disadvantage of this kind of cold box/fridge is that, when the boat heels over, the door can fly open and spill the contents.

✓ Hot air tends to rise and, as a result, top opening cold boxes tend to conserve the cold much better than front opening cold box/fridges. For the same reason evaporators must be installed as high as possible inside the cold box.

✓ Refrigeration units work best when the cold box is full; food and drink conserve the cold far better than air does.

✓ It does not really matter whether the cold box is switched on or off all year in terms of the lifetime and efficiency of the refrigeration unit. In any case it is recommended that you check to make sure your equipment is working well before the summer season starts. Just try getting hold of a refrigeration fitter once the season is under way...

✓ Although the output of a cooling unit will basically depend on the volume of the cold box, inefficient thermal insulation of the box itself will undermine the performance of the system. Minimum insulation thickness for a normal fridge is 5 or 6 cm, and this has to be increased to 8 or 10 cm for freezers. In both cases insulation thickness has to be increased at the bottom of the cold box, where the coldest air accumulates.

✓ Cold boxes are gluttons for electricity and whether or not you install one is going to have an enormous effect on the number and amperage of your batteries as well as your recharging system.

✓ The approach to operation and installation of freezers is practically the same as it is for cold boxes, while requiring an increased output to provide more cold. Ideally you should separate your cold box from your freezer, in order to avoid the loss of temperature that results from the constant and inevitable opening and closing of the cold box door.

Step by step

➤ *Before starting on the installation of the refrigeration unit we gave the cold box a thorough cleaning out and applied a couple of coats of white paint to an interior that had become dirty and deteriorated, with the old paint peeling off. However the original insulation was still in good condition and was sufficiently thick for our purposes.*

➤ *After finding and testing the model of evaporator and condenser, adequate for our 150-litre cold box, in situ, the fitter started to assemble the refrigeration unit by making a 20 mm diameter hole at a previously marked point in the top corner, where the evaporator plate was going to be fitted. This hole is for the gas pipes between compressor and evaporator, in Samba's case barely a palm apart. If necessary this separation can be up to three metres.*

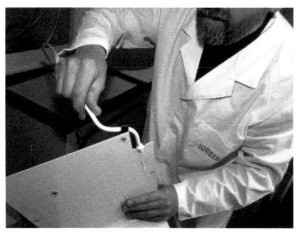

➤ One of the most delicate and difficult stages of fitting the cold box was the unbending of the spiral copper tube, so that it could be passed through the holes in the side of the cold box and the compressor housing. This operation, as with the following one of bending it back into its spiral shape, must be done with great care to avoid splitting the tube or obstructing it with too tight a curve.

➤ Little by little, and taking great care, the tube was passed through the small hole in the cold box. It really helps if somebody can lend you a hand with the operation of passing the tube through the hole.

➤ Once you have passed the tube through, you can attach it to the evaporator plate ('L' shaped model) using the corresponding screws. Remember that this plate must be installed as high as possible in the cold box, where the air in the compartment is at its warmest.

➤ Once the tube had been passed through the stern locker we once more fitted the insulation, to avoid condensation resulting from temperature changes.

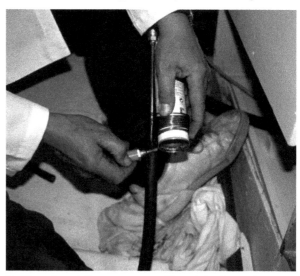

➤ To re-bend the copper capillary tube in such a way that looks suitably professional, a simple trick is to wind it round a can of spray paint or a tube of sealant, as if it were a reel.

➤ In order to lodge the compressor horizontally we screwed this shelf to the internal locker bulkhead. Four screws then attached the compressor in place on the shelf. The motor that we used came with its own silent-blocks and you can hardly hear it, even close up.

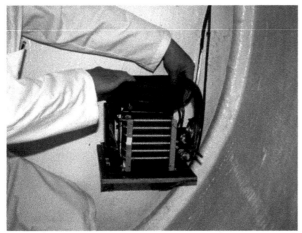

➤ The excess length of the evaporator tube, which was a lot in our case, as the compressor was so close to the cold box (barely 10 cm), must be coiled up with care and stacked on top of the compressor motor.

➤ Compressors are supplied full of refrigeration gas, which is released into the circuit by connecting the charging connections – sealed at the factory – to the tubes that lead to the evaporator plate. If, at any point, the compressor needs to be stripped down or disconnected this gas will be lost and the circuit will have to be refilled.

➤ When applying force to tighten the connections you can actually hear the moment when the gas is released into the refrigeration circuit from the compressor. It is important to control the correct tightness of these connections to avoid gas leaks.

➤ By the time we reached this point all we had to do was connect the thermostat that regulates the intensity of the cold in the cold box. A sensor was attached to the thermostat, which measures the temperature of the evaporator, with a wire passing on the necessary signals to the compressor, switching it on or off depending on the temperature in the cold box. We hid the sensor behind the evaporator plate. Using a couple of screws we then attached the thermostat, which operates in the same way as on any domestic refrigerator, in a visible, yet discreet, place.

➤ You can fill the hole around the tube and wire using special refrigeration putty (which is sold by specialist shops) which maintains its elasticity even when cold. This putty will avoid cold leaking out through the hole.

➤ When we had finished the installation we connected up to the power supply from the batteries (through the electrical switchboard) and the control wire from the thermostat. The work shown in this chapter took barely three hours to do. Part of the reason it could be done so quickly was the accessibility of the compressor in the stern locker. If the compressor is located somewhere more hidden away and difficult to access then the installation will take longer.

Chapter 29

Plumbing (part 1)

When the time comes to consider a freshwater system, there are endless possibilities, perhaps as many as there are boats, affecting everything from size, material used, quantity and location of the water tanks, and even the number and location of the taps. For each boat the system will be different, though the objective of each system is the same.

Pressurised freshwater system

All of *Samba*'s plumbing, and the same is probably true of many other boats of her age, was now obsolete, both in terms of comfort for her crew and from the point of view of safety in general. In the end, the only parts that we stuck with were the original stainless steel tanks (after we had given them a thorough interior clean up), completely replacing everything else.

Most of the work that you can see in the present chapter falls within the technical reach of most amateurs yet, while the complexity of replacing a water supply hose or installing a bilge pump is not in itself excessive (and for which you will not need any special tools), the consequences of any mistakes are quite drastic. When it comes to dealing with on-board water systems, whether freshwater or seawater is involved, you cannot afford to cut corners. The installation work has to be meticulous and you must always use the appropriate parts. If a clamp or through-hull lets you down, particularly when it is below the waterline, it can thoroughly spoil a day's sailing or even send your boat to the bottom in a matter of minutes.

To begin with, all boats are fitted with one or more tanks for the storage of fresh water. These tanks may be made of flexible plastic, fibreglass or stainless steel, and are usually fitted with a filling cap. They should also be fitted with a breather-hose so that the air will escape when it is being filled and enter as it is emptied.

Supply hoses then run from the tank, or tanks, to the pressure pump, which automatically switches on and supplies water when you open one of the taps. In the case of electrical pumps, a single motor is capable of moving water round the whole circuit. If hand or foot pumps are used (these are increasingly less frequent for freshwater systems), then each pump delivers water from the tank through its own pipe.

The general layout of a freshwater circuit is a relatively simple affair, although when considering a new installation it is best to set it out on paper noting the position of the tank/tanks, the pressure pump and the taps you want to install, as well as exactly where you want to run the hoses.

When drawing up this diagram you can also indicate the required pump flow rate, the lengths and diameters of the pipe you are going to use, the number and size of the clamps and possible connectors, the bypasses, filters and valves to be installed. Taking everything into account and calculating exactly where it has to go

can be quite complicated; there always seems to be something missing! But there will be no guarantee of success if you do not prepare the work meticulously beforehand.

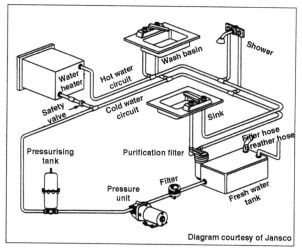

Diagram courtesy of Jansco

➤ *The above is the basic scheme for any pressurised freshwater installation. In the case of* Samba *we decided to do without hot water, although we did fit the taps in case we changed our minds at a later date. Before starting work we first had to design the installation, noting down the necessary numbers and sizes of the different valves, through-hulls, fittings, bypasses, clamps and inserts, not forgetting the diameter and lengths of hose that we would be using.*

Some loose ends

✓ One of the problems you will come up against when redesigning a pressurised system is in calculating the diameters of the hoses, threads, valves and fittings for the system as a whole. From the filler cap to the very last tap and including the pump, filters and bypasses, everything must be correctly connected: like fitting the pieces of a puzzle.

✓ With plumbing (nautical or domestic) you generally find the gauges for the threads and diameters are in inches. People who are used to a metric system will get completely lost unless they swot up on metric equivalents.

✓ You will find a wide range of products available for the different parts of your freshwater circuit. These are often exactly the same as the parts used for household plumbing or even systems for watering your garden, as well as all in your specialist nautical suppliers.

✓ For pressurised water and bilge systems you have to use reinforced hoses, which may be either spiral stainless steel wire or interior reinforced textile hoses. Avoid garden or transparent plastic hoses, because neither are capable of adequately dealing with the inevitable curves and corners (they will tend to kink and cut off the water) encountered when running the pipes through lockers and bilges, not to mention their tendency to split as they age if they are of inferior quality.

✓ The clamps must be made of stainless steel and, if possible, have a wide and bevelled base so that the pipes can be effectively squeezed without any danger of them shearing.

✓ Domestic taps will stand up quite well to marine conditions and when they fail (on average after around five to seven years), replacing them is neither too complicated nor too expensive.

✓ PVC valves, pipes and fittings that are frequently used in garden systems, identifiable because of their grey colour, should be avoided on boats as the salt atmosphere will weaken them in the medium term. PVC is also flammable and gives off toxic fumes when burnt.

✓ You can find through-hulls made of brass, stainless steel or fibreglass reinforced polypropylene. Brass is quite a cheap metal and reasonably well adapted to marine conditions but should be changed every five years or so. Stainless steel is longer lasting and, as a result, more expensive. Fibreglass reinforced polypropylene is generally black in colour and is usually chosen when one of your goals is to keep the weight down (it's a third of the weight of brass). The durability and resistance levels of this material, approved by Bureau Veritas, are very high, and it also has the advantage of being unaffected by electrolysis. However, it costs four times as much as brass through-hulls.

✓ For through-hull to hose connections, below the waterline it is best to use two clamps and not to tighten them all the way, so as to avoid splitting the plastic.

✓ It is a good idea to tie safety plugs off to their corresponding through-hull with a length of string. This means that, in an emergency, you will not have to grovel about in the bilges or at the bottoms of lockers to find them.

Step by step

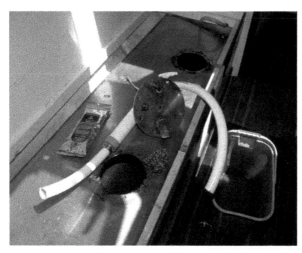

➤ We started the work of replacing the freshwater circuit by cleaning out the rust that had accumulated inside the tanks. First we opened the inspection hatches, which fortunately were quite big, and emptied out the dirty discoloured water that had accumulated inside them.

➤ Little by little, with the help of clean water, scourers and durable sponges, the tanks began to regain their original interior splendour. You should avoid the use of aggressive chemicals (descaling agents, acids, etc.) for interior cleaning of the tanks as any remnants could subsequently contaminate the water.

➤ We finished the work of restoring the tanks by changing the inspection hatch lid gaskets. Using the old gasket as a template we cut new 2 mm thick gaskets from a piece of nitrile rubber (resistant to water, diesel fuel and

most chemicals). To do this all you need are scissors and a punch. By the time we had finished, the tanks had recovered their original appearance. As a general rule it is a good idea to inspect the interior of your tank at least once a year. Once the tanks are clean you can start to think about how to disinfect the freshwater, either with a dilute solution of bleach (using only bleaches that are adequate for this purpose) or one or other of the drinking water products on the market.

➤ On most boats the tanks – when there are more than one – are connected at the bottom and fill simultaneously. A simple stopcock in the bilges is enough to avoid the water sloshing about from one tank to another while sailing. In the case of Samba, the tanks are filled from the top and a 'T' connector directs the water from the filler hose to both tanks. While this is a rather primitive system, it works and it would be a complicated job to redesign it.

 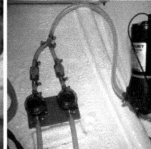

➤ The pressure unit that we installed came pre-assembled with its accumulator tank and is capable of supplying up to six taps simultaneously. To begin with we only installed three but it is always a good idea to over rather than underestimate your possible needs, paving the way for any extensions you may wish to make in the future. We also fitted filters to the supply pipes and stopcocks that allowed for the water to be transferred from one tank to the other, or to shut off one of the tanks in case of problems with the circuit. Everything was neatly organised and easily accessible underneath the washbasin in the head.

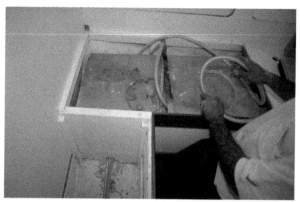

➤ Samba was originally fitted with a foot pump in the galley and an electric shower/tap in the head, with each circuit being independently supplied. We decided to redesign the whole system and layout of the hoses. One by one we ran the supply and pressure hoses (reinforced PVC) back to the corresponding points of supply. You should always try to find the most direct routes (involving the fewest curves) for the hoses, while also avoiding running them through the bottom of the bilges, where they will always be exposed to the accumulated dirt.

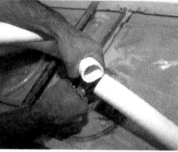

➤ When cutting stainless steel spiral reinforced hoses, start by cutting the plastic part using a metal saw or a very sharp knife. Once you have made this cut you can then use a pair of cutting pliers to snip through the metal wire flush with the line of the cut. If you try to pull or cut this wire any other way you could end up damaging the spiral or leaving part of the wire protruding out, risking future accidents.

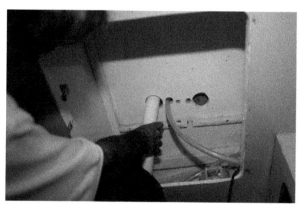

➤ We took advantage of the old holes in the bottom of the galley bulkhead to run both the starboard tank filling hose and the hose running from the pump to supply the galley freshwater tap.

➤ In the galley we fitted two taps, one for the freshwater and another for saltwater. The first of these was a domestic model, and the first step in the installation was to drill a hole in the counter top.

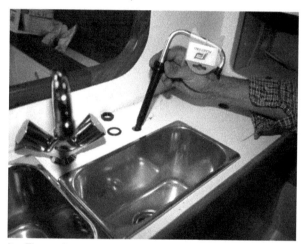

➤ The saltwater tap was to be supplied by the same foot pump that was originally used on Samba for the freshwater, now connected up to a through-hull located underneath the sinks.

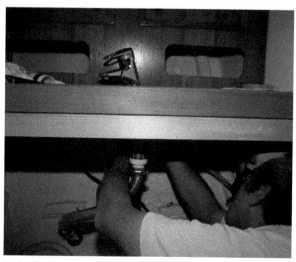

➤ In the head we installed a mixer tap and also a discreet pull-out shower. When we had fitted all the taps all that was left was to connect up the supply.

➤ Having a shower in the head is difficult to forego these days on any kind of cruising boat. After measuring up the cut, we drilled holes in the corners of the rectangle, into which we later inserted the jigsaw. A bit of sealant and six screws were enough to fit the shower in place, ready to be connected to the supply hoses.

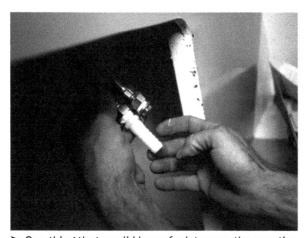

➤ One thing that we did leave for later was the question of hot water. However, all the taps, including the shower in the heads, were fitted with this future possibility in mind. The tap supply hoses are fitted with an insert to be fitted to the supply hose, and a threaded plug for the water outlet. All the connecting threads are sealed using Teflon tape to avoid drips.

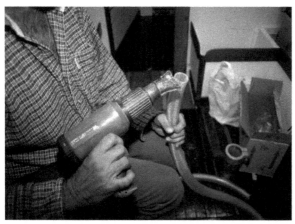

➤ It is not always possible to match the precise gauge of the hose to that of the terminals. In these cases it is always possible to increase the hose diameter a little. This

can be done by either submerging the end of the hose in very hot water or by carefully applying a hot air gun. Never try to expand the end of a hose by more than one or two millimetres in diameter. If you do, the plastic material and the interior reinforcing frame will be weakened and lose their consistency.

➤ The installation culminates with the installation of the breather hoses, which allow for air to enter and leave the tanks as the water level goes up or down. These should always be installed as high as possible and, preferably, on opposite sides of each tank. This will avoid water (or diesel in the case of the fuel tank) from spilling into the sea whenever the boat heels over.

➤ In this photo you can see our plumber making sure that everything is working correctly and that there are no circuit leaks or drips. It took two people two days to finish off the freshwater installation illustrated in this chapter. Yet another step forward in terms of on-board comfort.

Below deck a boat must be kept dry and free of water. An owner's first concern, before considering bilge systems, is to make it as difficult as possible for seawater to get in at all.

Bilge systems

The mere possibility of seawater sloshing about below deck is one of the worst nightmares any sailor can have. Every through-hull is a potential source of problems, as is the stern tube or the electronic transducer.

In principle all of these elements should work properly and so, ideally, the sea will respect the boundaries of the boat. But when this fails, as it sometimes does, our first line of defence in solving the problem and avoiding, or at least alleviating, any inopportune ingress of water is the bilge systems.

Safety regulations make it compulsory to install one or more bilge pumps, capable of bailing out 4,500 litres per hour in the higher sailing categories, and 1,800 litres per hour for smaller craft. These quantities of water may seem quite impressive at first sight but are ridiculous when you consider that if a normal size (32 cm) through-hull fails you could have almost 12,000 litres per hour pouring in. It does not matter how many pumps you have if the boat develops a serious leak; there is no way you are ever going to keep it afloat. With luck you might be able to alleviate the problem but never avoid it completely.

The purpose of the bilge pumps is to resolve small, chronic leaks (e.g. around the stern tube) or occasional problems (such as big waves, water from the shower, rain, etc.). If you have a serious problem (a broken through-hull, a hole in your hull, etc.) the best response is to try to repair or plug it, rather than sitting back and relying on your bilge pumps to keep you afloat.

This introduction, a little negative maybe, should serve to remind us all of the importance of making sure that seawater does not get in at all and, if it manages to do so, we can only hope that it is in a controlled way and in small quantities.

Bilge, discharges and heads

In this chapter we will be taking a look at the different bilge pumps installed aboard *Samba* and the set up for the different through-hulls, discharge systems and the heads installation. As has been the case with most of *Samba*'s new installations, the first step involves an important overall planning stage and preparatory work, which, in this case, was completed long before we actually started on the work shown in this chapter. The position of the bilge pumps, for example, was set when we designed the electrical layout and we had already installed the wiring. The carpenter then made the required holes through the furnishings and bulkheads, through which to run the discharge hoses. Without this preparatory work we would have ended up pointlessly wasting a lot more time installing our bilge system.

➤ *The hoses, valves, through-hulls and fittings, all of which will have to be fitted together.*

Bilge pumps: Different types and how they work

Basically there are three types of bilge pumps: manual, electric and engine-powered.

Manual pumps are necessary in the cockpit, installed in such a way that they can be operated by the helmsman.

Electric pumps can be divided into two groups; submersible pumps, with their motors installed in the bilges and with a body that also functions as a strum box; and self-priming pumps, that can be installed anywhere on board and are connected up to a strum box in the bilge and the discharge through-hull. Submersible pumps have the advantage of a simplified installation and can also run dry (when there is no water) without any great problem. The main attribute of self-priming pumps is that they can deal with particles without damaging the pump mechanism, as well as the medium term advantage of having the mechanism of the pump sitting high and dry, not in permanent contact with the water.

Engine-powered pumps, as their name implies, are driven by a belt that is connected up to the boat's main engine and are standard fittings in most large yachts. Apart from that, they function in pretty much the same way as self-priming pumps.

An alternative system to engine-powered pumps, and quite easy to set up in any boat, is to connect up a bypass with a stopcock to the refrigeration inlet motor and a suction hose running down to the bilge. In the case of a continuous entry of water, the normal motor inlet through the hull is closed and the bilge bypass connected. This means that the actual refrigeration system motor can be used as a powerful bilge pump.

Whatever the type and number of pumps installed, it is imperative that the water is discharged outside the boat by means of a through-hull, and never directly into the cockpit, regardless of how big the cockpit drainage holes are.

Another aspect to be taken into account is to fit the strum boxes as low as possible in the bilge, where the shipped water accumulates. Most boats have a gap for this purpose, usually notched in above the keel, where the different strum boxes or submersible pumps can be installed.

But theory and practice are two different things. In the case of *Samba*, for example, there was no notch for the pumps or strum boxes and, to complicate matters even more, there were two separate areas where water accumulated in her bilges; one just aft of the engine and the other above her keel.

Before designing *Samba*'s bilge pumping system we consulted a number of other North Wind 40 owners and each of them had resolved the problem in a different way. Some had chosen to join the two bilges together while others had completely isolated them and provided each with an independent pumping system. We finally decided to adopt this second solution.

Bilge systems: Some loose ends

✓ By screwing your strum boxes to the bottom of the hull you will ensure that they work more efficiently and will not move about. If you leave them loose you can change their position when the boat is heeled over, pumping out any water accumulated at the sides. Both systems have their advantages.

✓ Although, in the case of the pressurised freshwater installation it was possible to find adequate solutions for our needs in ordinary plumbing or gardening shops, in the case of bilge pumping units, the best alternatives are the specialised nautical brands.

✓ It will serve no purpose whatsoever to connect a bilge pump to a through-hull below the waterline, as you will not be able to generate sufficient pressure to discharge the water. The bilge pump discharge must be installed as high as possible in the topside.

✓ Through-hulls that are below the waterline must be designed for nautical use (brass, stainless steel or fibreglass) and must be fitted with their corresponding seacock. This, however, is not obligatory in the case of bilge discharge through-hulls that are installed higher in the hull which can be made of nylon or polyamide without causing any problems.

✓ It is a good idea to unify the diameter of the hoses that you are going to use for the bilge pump system. Apart from simplifying the question of supply, it will mean that, if any of your pumps or hoses fails, then it will be far easier to bodge together an alternative circuit.

✓ Regardless of how many manual or electric bilge pumping systems you install, you should always have a bucket, bailer and manual pump to hand in case of emergencies.

✓ All of the bilge pump hoses must be reinforced, incorporating a steel wire spiral or, at least, textile braiding.

Step by step

➤ You can start the work of setting up the bilge system at either end. In our case we started by locating the strum boxes in the lowest point of the bilges, where the water accumulates. It is important to have defined the location of pumps and through-hulls, as well as precisely where the hoses are going to run between them.

➤ The non-return valves stop the water from flowing back and improve the effectiveness of the pumps. They can be independent, incorporated to the strum boxes or coupled up to the through-hulls.

➤ For Samba, which has two seperate bilge pumping points, one in the centre of the boat and the other aft of the engine, we installed two self-priming electric pumps, the first just below the worktop in the head and the other in the stern locker, alongside the cold box compressor. One advantage of these kind of pumps is that they can be installed high and dry, away from the water and dirt that accumulates in the bilges.

➤ Section by section we ran the reinforced hose from the strum box to the pump and from the pump to the non-return valves and then on to the through-hulls, which we installed as high as possible in Samba's topsides.

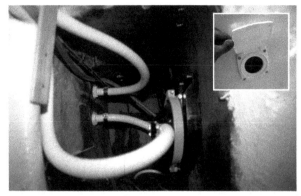

➤ In the cockpit we fitted a manual pump capable of moving 50 litres per minute. We installed the pump mechanism inside the stern locker, between the strum box and the discharge through-hull. In the cockpit the only part you can actually see is the lever, alongside the helmsman's position.

➤ To avoid drips we used Teflon tape to line the threads, and also rubber seals on the fittings wherever possible. Some shipyards have complete faith in polyurethane sealant for the watertightness of through-hulls, valves and fittings. This is a reliable and quick solution.

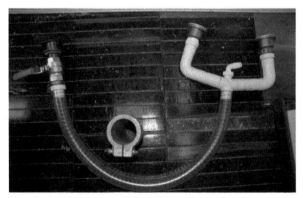

➤ This is the layout of the drain in the galley (the system in the heads is similar except for the double elbows under the drains). These components are common in production boats and, except for the profiled through-hull, the rest of these parts can be found without problem in ordinary plumbing suppliers. The photo does not include the clamps that, if you remember, should be profiled in stainless steel and installed two at a time when below the waterline. It is a good idea to lay out all of the parts and fittings in place, as shown here, for a visual check that everything is going to fit. The possible combination of through-hull, hose and fitting dimensions is vast but, fortunately, so is what is available.

➤ The toilet circuit, with its water input and outlet, is installed behind the cupboards in the heads. In this way hoses and valves are hidden from view but easily accessible for handling and inspection. To avoid a siphon effect, which may occasionally result in the entry of water through the toilet bowl, the installation of these hoses incorporates a large upwards loop. Whatever the case, shutting the valves after every time you use the toilet is recommended.

➤ To avoid having to screw the float switch onto the hull we glued it in place in the bilges alongside the bilge pump strum box, using polyurethane adhesive.

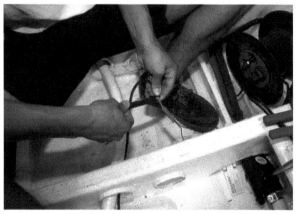

➤ You have to make sure that the electrical connections are all watertight. Apart from using quality connectors we lined the terminals with insulating tape and placed the unit as high as possible in the bilges, and then lined the assembly with self-vulcanising tape. You cannot take too many precautions in these wet areas of the hull.

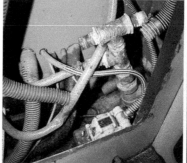 

➤ The two electric bilge pumps are controlled from the control panel above the chart table. They can both be operated manually or automatically.

➤ They say a picture is worth a thousand words and these two photos are a perfect illustration of the change below the sinks in the galley. On the left we have a veritable rat's nest of hoses and fittings that Samba had, somehow, managed to accumulate over the years up to her refit and, in contrast, on the right we have the neatness and clean lines of her new fitted drains and pressure hoses. The strange thing is that, apart from the evident improvement from an aesthetic point of view, the galley is now supplied with both freshwater under pressure and seawater by means of a foot pump, which is more than before.

Fitting a through-hull

➤ Uniformly pressing by hand enables the sealant to squeeze out all around the through-hull.

➤ After making a hole in the hull using a hole saw of the correct size, we started the installation by squeezing a bead of sealant round the interior bevel of the through-hull. If your interior through-hull fitting has an elbow that needs to be lined up pointing in a specific direction, the trick is to previously assemble the two parts (without sealant) and then mark the position of the through-hull in relation to the hull. By maintaining this reference the elbow can again be lined up in the direction you want.

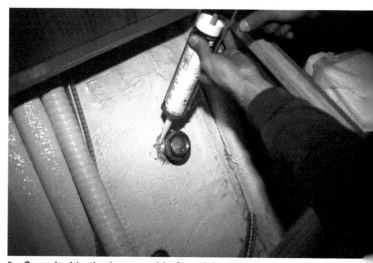

➤ Once you have squeezed the bead of sealant around the through-hull, carefully insert it into the hole.

➤ Once inside the boat, and before tightening the nut, squeeze a bead of sealant all around the bottom of the thread. Through-hulls need to be tightened firmly but not excessively, so as not to cancel out the effect of the sealant, which needs to be of a minimum thickness in order to do its job.

Chapter 31

Deck hardware (part 3)

Little by little the deck was now starting to come together. In this chapter we will look at a number of deck finishing and hardware solutions, all of which play their small part in the overall refit.

Advancing on various fronts

In previous chapters we showed you how we installed a good part of *Samba*'s new deck hardware and rigging, however there was still a lot of important work to be done before we completed our refit and brought the deck completely up to date. Some of these jobs are barely worth mentioning, such as the reinstallation of the original fittings with four new bolts, but there is a small selection of jobs that we would like to take a closer look at. In some cases this is because of the ingenuity involved, in others because they are quite complex and, finally,

in order to justify technical decisions that may serve as an example for others.

Synthetic anti-slip in the cockpit

On fibreglass boats the passage of time, endless sluicing down plus the simple toing and froing of the crew results in the inevitable, if gradual, wearing away of the deck's moulded anti-slip pattern. When you add to that the occasional knocks received and the peeling and flaking suffered by the deck as a whole then you are faced with a situation, both inevitable and typical, where the need for restoration is imperative.

The solution most frequently resorted to for restoring some of a boat's original look is a coat of paint. This is indeed the first thing we did, although we knew that the paint was going to clog up the moulded anti-slip pattern, smoothing it out and reducing its effectiveness. There are different solutions to this inevitable problem. One of these is to line your deck with teak, which is a very seductive finish but one that will add a lot of extra weight to your boat, while removing it from your pocket! In the case of *Samba* we decided on a different solution. On most of the deck we applied a final coat of special non-slip paint, but in the area of the cockpit we opted for an adhesive polyurethane covering.

Synthetic coverings are pretty much standard for deck refits, as well as being a solution used by many shipyards for their new boats. Depending on the manufacturer, these composite coverings are sold in rolls of different thicknesses, colours and patterns. There are even coverings in strips that simulate teak planking, in a wide range of colours. These coverings represent a simple and quick anti-slip application system (approximately 1 m^2 per hour), offering a good grip and weather resistance at a reasonable price. This was the solution that we settled on for *Samba*'s cockpit, and the first part of this chapter deals with the installation.

Step by step

Synthetic anti-slip in the cockpit

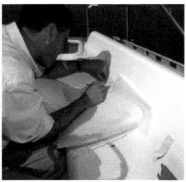

➤ *Before fitting the synthetic coverings you will have to prepare the areas where they are going to go and sand them down with fine-grain sandpaper. These areas then have to be thoroughly cleaned and degreased. Use tape to mark out the edges of the areas you are going to cover, although you can also do this with a pencil or marker, and then make up templates using tracing paper for each of the pieces you want to apply.*

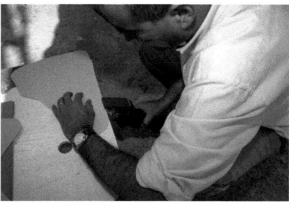

➤ *Once each piece has been trimmed and adjusted you should then sand down the edges all the way round to reduce their thickness. Apart from improving the way it looks when installed, reducing the profile will also help to avoid it being yanked off accidentally due to a stumble or brusque manoeuvre. You can do this sanding either manually or with a power sander.*

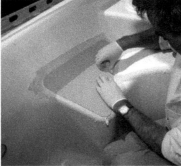

➤ *Once you have these templates it is easy to mark out the section of coating that you need for each area, whether in straight lines or curves, and all you need then is a cutter and/or scissors. As you cut the sections out it is a good idea to fit each one in place and make any adjustments that may be necessary to ensure that they have the precise shape required.*

➤ *On the helmsman's bench, which still retains some of the original anti-slip pattern, we glued the covering on using contact adhesive, first of all applying generous quantities of adhesive to the surface.*

➤ We then used a 1 mm toothed spatula to spread the adhesive uniformly over the whole surface, avoiding lumps or any excessive thickness that could form lumps underneath the coating. If you mark out the edges of the part using tape, instead of pencil or a marker, this will simplify the problem of the adhesive getting squeezed out and later having to clean it up.

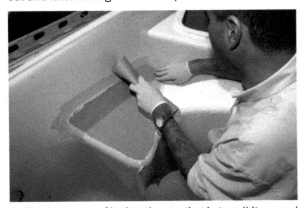

➤ The best way of laying the coating is to roll it up and then gradually unroll it over the surface, pressing down gently with your hand to avoid any air bubbles getting trapped. When the coating has been fitted in place and before the adhesive has had time to set, you can still make slight adjustments to its position.

➤ Finally, using a rubber roller and without applying excessive force, in order not to squeeze out the required thickness of adhesive, run the roller over the whole surface to make sure that the section is uniformly glued. Polyurethane is a contact adhesive, which means that you do not need to apply pressure or weights to the glued parts. All that remains to do after this is to clean away the excess adhesive that gets squeezed out round the edges. Excess adhesive that has not set can be eliminated easily using paper towels soaked in soapy water.

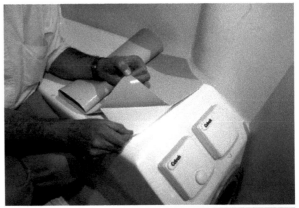

➤ For the cockpit benches we decided on the self-adhesive version of the polyurethane coating, which is particularly effective on flat surfaces. The process for drawing out and then cutting the pieces is identical to the one explained previously. The only difference is the way that it is glued to the surface. With self-adhesive coatings it is better to start with it rolled up. In this case you have to make sure that the piece is correctly lined up from the start. Once it has stuck to the surface, even if only partially, the firmness of the adhesion is such that it is impossible to move it without ripping the paint off (in our case), or spoiling the coating itself.

➤ Although the application of self-adhesive coverings requires a more precise technique than gluing it with contact adhesive, it is much quicker to do and far less messy. Self-adhesive pieces are also ready to use as soon as they have been fitted in place.

➤ A common problem with the installation of synthetic anti-slip coverings is trapped air bubbles. Apart from doing your best to avoid them, by attaching each piece with great care, progressively, should you find an occasional air bubble there is no need to panic or stop what you are doing: once the section is in place simply use a cutter to prick the bubble and it will disappear without leaving a trace.

➤ Perhaps the new cockpit synthetic anti-slip covering is not the most luxurious finish that you are going to find on a boat, but it is functional, economic, hard-wearing, easy to install and maintain and also complies perfectly with its essential purpose, which is to avoid your feet slipping away from under you.

Installing the mainsail traveller track

The installation of mainsail traveller tracks on cabin roofs has become a standard feature on cruising yachts but, from a technical point of view, this presents a lot of problems. On the one hand, it distances the car and traveller from the wheel, which means that you have to move around the cockpit every time you want to adjust the sail. On the other, by moving the sail attachment forward, the stresses involved in hauling, and those sustained by the boom, are almost doubled, adding unnecessary resistance and risk of breakage. We also have to add that many of the cars that we have seen installed on cabin roofs can barely manage to luff the mainsail, as their lengths are truly ridiculous. If you then add the fact that the sizes of the mainsails on modern yachts have been progressively increasing, in comparison with the genoas, we cannot do more than reiterate our opinion regarding the inconvenience of this layout.

Given this conclusive criticism of locating the mainsail traveller on the cabin roof, many of you may be wondering why, in the case of *Samba*, that is exactly where we decided to put it. This was not an easy decision to make. On the one hand we had all the arguments, outlined above, while on the other side of the scale there were three factors that weighed decisively in our decision. The first was the reduced size of *Samba*'s cockpit; even after we had moved the traveller track out of the cockpit, the benches were still only about a metre in length, which is laughable for a 40-footer. At the same time her owner (who happens to be the author of this book) many years ago had a motorbike accident, one of the results of which was reduced mobility in his left knee, which meant that avoiding functional barriers in the cockpit was of primary importance.

Finally, we must also add that the North Wind 40's mainsail is barely 24 m², a size that these days you are more likely to find on a 28- or 30-footer, given that a modern 40-footer with its fractional rig can easily clock in at twice this area. As with most boats of her age (fully rigged), *Samba*'s main pull force comes from her forward sails, with genoas that can present up to 60 m², while the force exercised by her mainsail is quite limited.

In any case, if the solution adopted turns out not to be satisfactory, bringing the traveller back into the cockpit would just be a question of a few hours' work. The holes for attaching bolts and reinforcements are still in place under the cockpit benches.

➤ In order to install the traveller track on the cabin roof we designed double steel brackets to hold both ends of the traveller track. These two parts were attached to the deck using six bolts.

➤ In this view from the interior you can see the plywood reinforcement, which we laminated to the underside of the cabin roof in anticipation of the new layout. We also fitted a stainless steel plate that would help to distribute the stresses and provide the assembly with greater strength.

➤ When we had the brackets firmly attached we installed the traveller track, a self-supporting model (that supports the stresses with the central part in flight) and the attachment bolts of which are integrated and hidden in the profile of the track.

➤ The layout of the rigging on the cabin roof allows the halyards to be passed through the brackets supporting the traveller as they run from the mast, arriving in a free and orderly manner to the rope clutches and their respective winch.

Side-mounted cleats: applied ingenuity

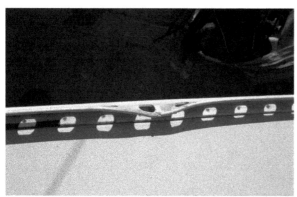

➤ One of Samba's multiple war wounds was a hard blow suffered by the toerail, about ten feet from the bow on the starboard side. Fixing this kind of damage is much more complicated than it may seem. Simply bending the toerail back into place is practically impossible. Apart from the immense force that would be required and the consequent risk of damaging the bolts or the fibreglass structure of the deck, the aluminium does not take kindly to being bent when cold and tends to break. Removing the whole length of toerail would also be a hard job, as the bolts are actually laminated to the hull and removing them one by one would be mind-numbingly tedious and laborious.

➤ In the end we came up with a simple and rather ingenious solution (involving about two hours' work). Using the grinder we cut out the damaged section of the toerail, rounded off the edges and installed a cleat bracket in the gap. The result, as you can see looks pretty neat and is also practical. The only drawback was that, to maintain symmetry, we had to do the same on the port side. Fixed in place with bolts attaching them to the base of the toerail these will doubtless be the most firmly anchored cleats on the boat.

Stanchions and lifelines

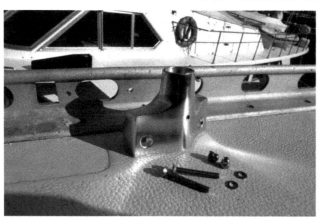

➤ Bent stanchions, with broken bases and lifeline cables that are exposed and rusted are a common enough sight among boats of a certain age, and Samba was no exception. A complete refit was evidently required. In our case we took advantage of the occasion (of painting her deck) to fill in all the holes left by the original stanchion bases, in the knowledge that every hole on the deck is a potential source for the ingress of water. In place of the original stanchion bases we installed new cast aluminium bases which we mounted directly on the toerail.

➤ Lifelines, stanchions and their bases make up an assembly that is designed to give way in a scaled manner when subject to certain forces. In the case of such accidents, the first parts to give way, as all boat owners are aware, are the stanchions, which will bend. Should the force continue, then the stanchion bases will give way, acting as a kind of fuse, and avoiding damage to the hull or deck. The strongest parts of this assembly are the lifelines, which ought to support a pull of around one ton before breaking. When the scale is as above, the safety of the boat and its occupants is paramount and the replacement of damaged elements is relatively straightforward. If the lifeline breaks before it should, this will put the crew in danger and, if the base of the stanchion holds out too long or the stanchions do not bend when subject to heavy loads, this could result in the delamination of the deck, which would be much more difficult and expensive to repair.

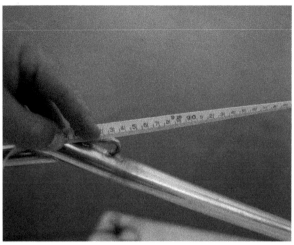

➤ Measuring the length of your lifelines is as easy as using a tape-measure and measuring the distance between the becket at one end and the becket at the other. First of all make sure that you measure the distance between the centres of the pins and not between the exteriors of the supporting eyebolts. The second recommendation is not to trust in the symmetry of your bow and stern pulpits. The lifelines of very few are identical in length and it is almost always the case that there are small but subtle differences in each of the four lifeline lengths.

Repair or replace: mooring chocks and backing plates

When you are refitting a boat it is often difficult to decide whether it is better to repair a part or just replace it. The case of *Samba*'s mooring chocks and backing plates – both very much deteriorated – serves to exemplify this dilemma.

➤ They stopped making these old bow mooring chocks years ago and finding a new model that was more or less adaptable (aesthetically and functionally) was far from easy. The simplest and cheapest option was to just buff them up and anodise them again. Salvaging your old parts will also save you all the work of drilling new holes and/or adapting the deck.

➤ For the backing plates, on the other hand, the solution was far simpler: a case of making up new ones using the old ones as a template. This was a much brighter and a far cheaper solution than wasting time trying to restore them to a pristine state.

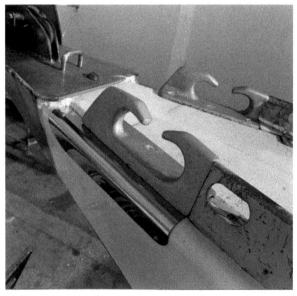

➤ The choice of repair or replace is not always as clear cut as this; in fact it is a constant dilemma from beginning to end of the refit. Once the 'new' re-anodised mooring chocks and their shiny protections had been fitted the imperfections of her gunwale stood out far more; another of the recurrent problems with refits. Everything you fix or replace tends to highlight the defects of the parts around it.

The mast itself was going to add little to our refit. To begin with it was simply a matter of restepping Samba's mast, but before we did so we took advantage of having the mast down to bring her electrical, electronic and rigging systems up to date.

Ready to raise the mast

There are few things that identify a sailing boat as much as seeing her with her mast up. With her mast raised, the whole of her deck has an air of completeness, a condition that we were, as it happens, gradually beginning to achieve in the case of *Samba*.

At the start of the refit we decided to stick with the original mast as the extrusion was still in an acceptable state. Then, after fixing and replacing various aspects of the rigging, we postponed a number of other jobs, such as replacing the halyards or renewing the electrical systems, scheduling these jobs for when the time came to restep the mast.

The work of bringing the mast's electrical and electronic systems up to date is always easier when you have it lying horizontally on dry land. Replacing halyards, on the other hand, is probably easier to do after the mast has been restepped.

Navigation lights
Some boats have their position lights (green and red) on the bow pulpit, while the stern light (white) is often mounted on the stern pulpit. Others, as was the case with *Samba*, had, and continue to have, these three lights mounted at the top of the mast, a positioning that we believe is more advantageous. Mast top lights are protected from sea-spray and can be seen from further away. At night the lights do not dazzle the helmsman and will light up the Windex, which is a big help. From an energy point of view, combined mast top lamps use less power, relying as they do on a single bulb

for the three functions (green, red and white). Their main inconvenience is the added weight of the wiring that has to be run up inside the mast, plus the fact that you will have to climb the mast every time you need to change the bulb. In this sense, and as soon as we recovered from the refitting, we intended to install an LED system, which is much longer lasting and consumes even less electricity.

In terms of the engine navigation light, for some reason that we are unaware of, *Samba* did not have one. As a result we took advantage of the opportunity to mount a unit that incorporated both an engine and a deck light.

Step by step

Bringing the electrical installation up to date

➤ We started the navigation/anchor lights module installation by attaching the base to the stainless steel U-bracket that was already fixed to the top of the mast for the riding light. We then ran a new wire up the mast, attaching it to the old one to pull it through. The wire was later passed through a rubber guide/seal (supplied with the deck light) to reinforce the watertightness of the set-up.

➤ As with all electrical equipment in the open air, protecting the connections is crucial to avoid rusting. Before attaching the pressure connectors, the ends of each cable were given a coat of tin and then wound round with self-vulcanising tape or thermo-retractable sheaths. Rather than being technically complex it is more a matter of doing this work meticulously.

➤ The combination of lights, seen here from aft, includes an anchor light below, with its own light bulb, and above the three sectors: red, green and white (stern). We also took advantage of the situation to install the wire and the base of the VHF aerial, as well as the base support for the electronic wind gauge, already connected up to its corresponding multiwire cable, which is also run down inside the mast. To avoid untimely knocks when stepping the mast, the VHF aerial, Windex and electronic wind vane were not finally installed until the mast had actually been stepped.

➤ **Wiring scheme:** In order for a bulb, or any other 12 V DC instrument to work the current has to 'arrive' through the positive pole and then 'return' to the negative. The route of arrival must be different for each light, each controlled by its own switch, but the return can be shared by more than one apparatus. This approach allows for a saving in wires, weight and, of course, money. As can be seen in this diagram, on the mast top tricolour and anchor light, and on the combined masthead/deck light unit, we used three-phase wire. Of the three wires in each conductor, two were used to connect up the positive terminals of each unit, while the third combined the two negatives. To avoid confusion when connecting these wires up to the control panel you should take a note of the colour used for each terminal – it is better not to just trust your memory.

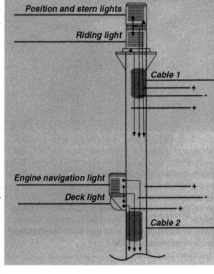

Masthead and deck lights

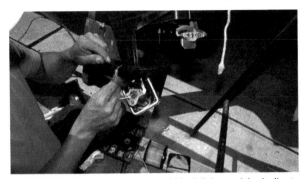

➤ Fitting the combined masthead/deck lights unit is similar to the system used for the mast top lights, although in this case it is easier to first connect and tape the cables into place before attaching the unit to the mast. The negative terminals are connected together and then joined to one single wire that runs down to the base of the mast and from there to the control panel (see the wiring connections diagram opposite).

➤ Four rivets were used to attach this light unit to the mast. The unit's system of adjustable plastic wings allowed it to be adapted to any mast size or profile.

Stepping the mast

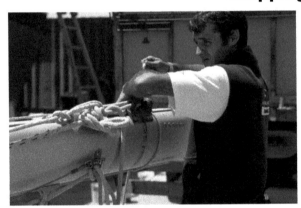

➤ Before thinking about lifting the mast with the crane you have to firmly attach all of the halyards to the mast, along with the running rigging, so that nothing is left hanging loose, with the risk of it getting snagged up during the lift. To avoid knocking or scratching the hull or deck it is also a good idea to wrap bits of carpeting around the ends of the shrouds.

➤ The ends of the spreaders have to be protected to prevent them being attacked by electrolysis due to contact between the aluminium that they are made of and the stainless steel of the shrouds. The best solution is to use insulating paste and then some kind of protection and/or sheave to protect the sails from rubbing against the spreaders.

➤ Before lifting the mast it is always good to take from advice the crane operator. The best point to take the weight of the mast, in order to lift it, is a little above its centre of gravity (1 or 2 metres). If you lift it from higher up you will find it is harder to line up the base for stepping, plus the crane may not be tall enough to lift the mast to a sufficient height. If you try to lift it from anywhere below its centre of gravity, disaster beckons, as the top end will just come crashing down under its own weight. When lifting the mast, attach a retrieval line to the lifting loop so that, when the mast is stepped, everything can be detached on deck rather than using the bosun's chair.

➤ It is recommended that there should be at least two people, apart from the crane driver, to control the movements of the mast as it is lifted. One of these must make sure the profile does not get dragged across the ground, while the other stands on deck and guides the crane driver in lining the mast up so that it can be stepped. Before doing the lift, make sure that mast, boat and crane are all correctly aligned. Turning the mast round after it has been lifted is extremely complicated.

➤ With the mast stepped, but still suspended from the crane, the time is right to attach the stays and shrouds with their terminal fittings to their respective shackles or jaws, the safety pins of which must always face in towards the centre of the boat or the stern (depending on the situation) to avoid getting snagged in sails or sheets. A little marine grease will ease the opening of the turnbuckles while also protecting their threads from saltwater aggression.

High performance halyards: a solution that saves weight

➤ With the mast once again in place we renewed the old halyards with new ones with a Dyneema (Spectra variant) core. To make this change you have to tie the ends of old and new halyards together using string, tying them off with various loops, which are then taped. If you suspect that this may be a tight fit as the join passes through the different sheaves, then it is better if the ends are firmly tied and taped together. Once this is done, all you have to do is pull on the loose end of the old halyard until the new one appears in its place. In less than an hour we changed all the yacht's halyards and lifts.

➤ Changing the halyards allowed us to observe the weight reduction that high performance materials deliver. The weight on the scales of the new high performance halyards was 13 kilos and, at the end of the operation, the weight of Samba's original, old, thicker halyards was 31 kilos. It is as if we had removed an 18 litre drum of water that had been permanently hanging from her spreaders. With this change alone any boat would improve her performance and sailing ability. It is true that, for the same thickness, high performance materials are significantly more expensive than polyester but, when you make the change, the diameters required can always be reduced by one or two sizes, cancelling out most of the difference in price.

Connecting up the mast

➤ As a mast fairlead we chose this curved stainless steel tube, through which the ropes can pass freely forming a kind of siphon so that the water will be squeezed out of them. Installing a standard fairleader with pressurised rubber would have meant making four more holes in the deck. These types of fairleaders also tend to form a cluster of ropes alongside the mast where it is easy for them to become accidentally snagged.

➤ Below decks, our electrician connected the wires running down the mast up to the circuit through two strips, one for the wind equipment and the other for the lights. The VHF aerial cable had to be fitted through a special connection. A bad aerial connection is the origin of almost all radio reception problems.

Considered twenty-five years ago as an extravagance for lazy sailors, these days a genoa furler is virtually indispensable on any cruising yacht. Over the last quarter century there have also been constant improvements in their design.

Installing a furler

These days few boats reach their customers without being fitted with a genoa furler. The advantages of a furler far outweigh the disadvantages, simplifying the handling of the genoa to such an extent that practically all leisure sailors are now convinced that they would find it difficult to manage without one. All genoa furlers consist of a profile around which the sail is furled, by means of a drum that is attached firmly to that profile. A rope wound around the drum makes this unit turn around the forestay.

The operating principle could not be simpler but what is required is that the furler turn smoothly and easily, without the need for force, and that the sail keep an acceptable shape. With the sail fully spread, the furler will barely interfere with its shape and should it fail to perform adequately this will be due to an inadequate trim or cut.

Things get more complicated when you start to furl. In this case the profile must, uniformly along its whole length, resist the twisting force produced by the partially furled sail. This is when the combination of a good cut and maximum profile resistance must come together to achieve the best performance.

But do not expect too much: after a certain point in the furling, around $^1/_3$ or $^1/_2$ of the sail area, the shape that this presents to the wind or the height of its pressure centre will inevitably reduce the performance of any boat, listing too far over or losing their upwind angle. The foresail rigging and/or shifting stays are the best option in these cases, but this is not the question that concerns us here.

Getting back to the furler, in order to better resist the torsion and guarantee a regular rotation, without jerking, the most effective shapes for the profile are undoubtedly rounded ones: a ball will always roll better than an egg. On the other hand, those who combine cruising with racing would probably choose an oval profile, presenting a more effective laminar flow at the sail's attacking edge or luff. Racing sailors also like to use drums that can be dismounted, drums incorporated in the anchor locker or extra-flat models with a continuous rope.

You will find that just about all of the furlers on the market work well, particularly in their use for ordinary leisure sailing. The differences in quality and price between different makes is basically a question of their lightness and the resistance offered by their profiles, the quality and size of the drum bearings and aspects of the finish that will contribute to the assembly's durability and ease of use.

Not all boats need a top of the range furler capable of handling four non-stop round the world trips although, at the same time, neither would there be much sense in installing the cheapest furler available, which could prove insufficient for our fairly modest requirements. If in doubt, the best thing to do is consult an experienced professional, someone who knows the strengths and weaknesses of each make and can advise you on the best choice for your particular boat, budget and the type of sailing you plan to do. It is also worth saying that many of the problems encountered with furlers are related to their being under-specified, incorrectly installed or badly operated.

By scrupulously following the manufacturer's assembly instructions, most owners should be able to install their own furler in a couple of hours or so, without coming up against too many technical complications. In some cases, getting rid of the forestay is advisable or even obligatory, while in others you may have to adapt a terminal to the stay in order to correctly anchor the drum. Apart from these details, assembling a furler is pretty much the same for most of the different makes and models on the market. It is a simple matter of putting together the different parts that make up the profile and then attaching it to the drum and to the boat.

Most furlers can be installed with the mast raised, although it is usually easier to do so with the mast on dry land. This was how we wanted to do it ourselves, but a technical problem (with the forestay turnbuckle when raising the mast) required us to assemble the furler on dry land, on a new stay, and then lift the furler and stay onto the boat using the crane. The processes are actually the same in either case, the only difference being that, with the stay on the mast, the parts have to be raised, either by hand or using the halyards, as they are assembled on deck.

Furlers: loose ends and advice on how to use them

✓ Friction bushings support the transversal forces of the furler's profile round the stay. In an ideal situation these bushings would be continuous along the length of the profile, although this is impossible due to the slightly curved forms that your forestay will always adopt when sailing. Each make, in its own way, resolves the problem of the best position, ideal number and type of bushings included in the profiles of their furlers.

✓ When installing the furler, and before disconnecting the stay, it is important to make fast a couple of halyards to the bow, in such a way that the mast will be firmly anchored and will not lean back too far towards the stern.

✓ To ensure the correct operation and long life of the furler you should always unfurl the sail slowly, using the control rope as a brake, which should always be kept tight, so that it does not become kinked when furling. You should never furl a sail that is fully drawn or fluttering.

✓ Do not force a furler that will not turn freely. This will often be due to a halyard wrapped around the stay. You have to keep an eye on your loose fore halyards (spinnaker, genoa, etc.) to make sure that they are clear when furling.

✓ If you use a sail that is much shorter than the length of the stay (not recommended) it is better to keep the halyard swivel as high as possible. To this end the simplest thing is to install an extension (rope or wire) to the head of the sail.

✓ Furlers hardly need to be maintained at all, except as indicated by the manufacturer, and, at most, applying a little spray-on lubricant. Maintenance is reduced in most cases to periodic sluicing down with fresh water.

✓ Increasing or reducing forward sail surface means relocating the genoa traveller car.

✓ Do not ease the genoa halyard while working the furler.

✓ Correctly hauled forestay and genoa halyard are indispensable if you want you furler to work properly.

✓ It is a good idea to mount the drum rope so that it will furl the sail in the same direction that the metal wires of the stay are wound. Turn direction will also have to be borne in mind when it comes to ordering a new sail, to ensure that the UV-protected side is going to be furled on the outside.

✓ The use of the furler does not mean that you will not need to keep a storm jib on board. There are a number of systems on the market for bending this sail onto the furled genoa. The ones that we know of are pretty tricky to set up, particularly bearing in mind that you are unlikely to be installing your storm jib in placid conditions. The idea of tying it off more or less like a staysail or a flying stay is certainly the most recommended option, with the result that you can avoid having to lower the genoa every time you get your storm sails out.

✓ To simplify the handling of a furler, use quality blocks for the control rope and make sure that you install them at the correct angle of attack, especially at the start (90° to the drum) and end of their run.

Step by step

➤ The first part that we threaded onto the stay was the deflector wheel, the purpose of which is to avoid the genoa halyard wrapping around the stay and blocking the unit. The halyard swivel was then fitted, with two special friction bushings (recognisable by their red colour) and finally locked in place using a pin.

➤ We then threaded on the plastic top cap and the top section, which was the only part that came without drill holes at the top end. Mounting the furler with the mast stepped is the same process as with the mast on dry land. The only difference is that, with the mast stepped, you have to push the parts up the stay, or pull them with the aid of a halyard. If you use a halyard, do not forget to attach a release-pull, so that it can be loosened when you have finished the assembly.

➤ Before we made up the stop we inserted two pairs of friction bushings into the standard profile section. The plastic stop was locked firmly onto the profile by means of a ribbed aluminium rivet, that we inserted using a mallet.

➤ The assembly sequences are repeated from this point on until you have put together the whole length of the profile. First we inserted an interior connector. This connector had two threaded holes which we had to line up with those of the profile section.

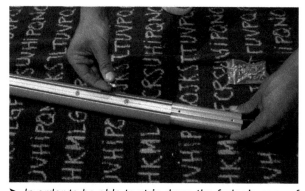

➤ In order to be able to strip down the furler in case of need (and also to avoid the screws coming loose over time) we recommend that you bathe the end of the screws in an electrolysis insulation product, although the furler screws we had were delivered with a coating that served this purpose.

➤ Before we attached each of the external profile sections we inserted a friction bushing.

➤ We then fitted the standard profile section that slides over the interior connector and again with the lower screw holes. Once this unit had been assembled uniformly we tightened the four screws.

➤ Once we had made up the full length of the furler we inserted the final part, which did not have front-mounted screws. Instead there were two countersunk holes in the sides where rivets could be fitted, in order to leave the furler at the required height.

➤ Inside this part we inserted a connector that was longer than the previous ones (1.64 m). This telescopic system allowed the length of the furler to be millimetrically adjusted without the need to cut any of the parts to size. At this point we also slid the plastic feeder wedge, which functions as a guide for the sails, over the stay.

➤ Before fitting the drum we then had to slide on the halyard swivel, making sure that we got it the right way up.

➤ Here we started to assemble the drum, the top part of which was inserted into the profile. An Allen screw locks the profile firmly to the drum so that the whole unit will turn as one part. The first stage of this operation had now been completed; it took us about an hour to do.

➤ A technical problem with the forestay turnbuckle, detected when stepping the mast, required us to assemble the furler on dry land, working with a new stay on which we adapted a Norseman swage terminal.

➤ Climbing a mast when your boat is on dry land is an extremely dangerous thing to do and most shipyards will not allow it. The stability of a boat on her berth is much more precarious than it is in the water and could easily overbalance as a result of the leverage exercised by the weight of a person at the top of the mast. For this reason we again hired the crane to raise the stay/furler assembly into position.

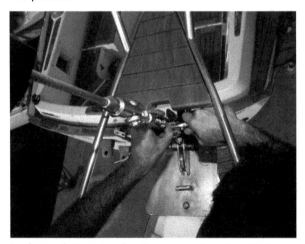

➤ Some furlers require the stay to be adapted. In our case the drum could be anchored using the boat's existing turnbuckles and pad-eyes or even combining the two systems. Once the turnbuckle had been adjusted to exert the correct pull on the mast, it is locked into place by a locking pin.

➤ With the turnbuckle adjusted it was time to attach the drum. We did this using special stainless steel link-plates (these are supplied with the furler) which are then attached to the stay mooring pad-eye.

➤ After this we raised the profile of the furler until it came up against the top end of the stay and then allowed it to drop 5 cm, giving us the height recommended by the factory for attaching the profile, as a whole. Through the holes in the exterior profile, and taking care not to damage the stay, we drilled a 3 mm hole on each side of the telescopic section.

➤ Two rivets were then used to firmly fix the two parts of the telescopic section. These rivets (supplied with the furler) appear to be quite normal but, when you take a closer look at them, you can see that they are countersunk, which means that they will not stand out beyond the surface of the profile and interfere with the raising or lowering of the halyard swivel. Finally, we firmly fixed the sail guide in place, also with two rivets, bringing the installation of the furler to an end. In total, including lifting it up to the masthead with the crane, the whole job only took us four hours.

Chapter 34

Interior carpentry (part 4)

The way things are these days, the cost of skilled labour is going to be the most expensive part of any repair, whether you are refitting a boat, taking your car to the garage or calling in a plumber to fix a burst pipe.

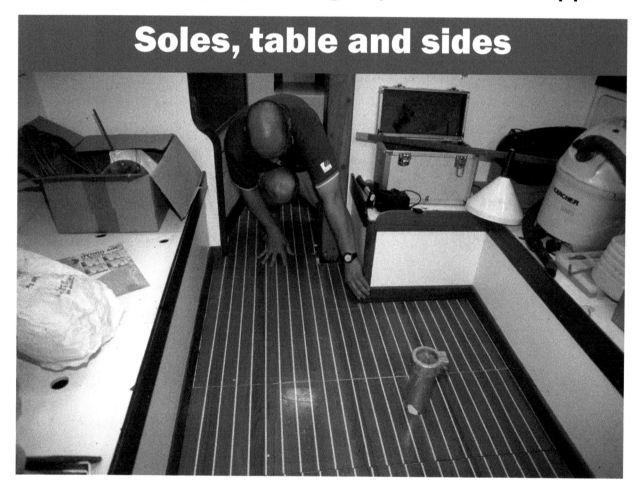

Soles, table and sides

Renovation, restoration and repair are now the work of craftsmen, while modern tradesmen have become experts in replacing parts. Which you use is purely a question of economics because, at the end of the day, it is usually going to cost you more to repair something than replace it, and the high cost of skilled labour often excludes repair as a reasonable option.

Bearing this in mind we considered the restoration of *Samba*'s cabin soles, saloon table and the walls of a number of below deck spaces. To employ the analogy

of a football score you could say that replacement beat repairs by two goals to one. But let's take it step by step.

Firstly, the soles: *Samba*'s original soles consisted of plywood boards with 3 mm thick teak strips glued onto them. Over time these had become unstuck and some of these strips had got lost and had to be replaced. To repair the soles we would also have had to strip down the surface of the teak strips, always hard work, and revarnish them. At this point, to continue our football analogy, we still had a goal-less draw on our hands but, in the end, the die was finally cast in favour of replacement (1-0). This was due to the fact that, at the entry to the cabin, who knows for what reason, the sole simply consisted of a plywood board without any teak finishing strips or anything. Replacing this area and getting it to look the same as the rest would have been expensive and hard work.

We decided on a solution based on marine plywood boards lined with a synthetic high-resistance panel, which was really easy to install. All we had to do was trace out the shapes of the original soles on the new boards, cut the new boards to size and fit them into place. Although these boards were quite expensive, the solution paid for itself due to the saving in skilled labour costs.

If the original plywood boards had been in better condition then we would only have needed to buy the synthetic veneer, which we could have glued onto the original sole boards. But, even if this had been possible, it is not certain that the saving in raw materials would have compensated for the extra work involved.

The side walls of the saloon were originally fitted with pine tongue and groove planking. You can find this type of planking in just about any DIY store, with an infinite range of finishes and colours, at a very reasonable price. Stripping down, repairing and varnishing the old planking is simply throwing money down the drain. It is far easier to buy new planks, which is exactly what we did.

With replacement now having a two-nil lead, restoration managed to pull one back when it came to the saloon table. *Samba*'s original fold-away double-wing table was a good example of the cabinet-maker's art, with a solid teak structure and raised-rim mouldings. Furthermore, the size of the table when unfolded (1.15 x 1.10 m) was just right for a 40-footer. The biggest problem was the veneer, which was badly scraped in some areas and peeling off, and had also been blackened by damp in others. Once we had removed this veneer, instead of replacing it with the same we decided on a synthetic finish, perfect for a surface like the saloon table, exposed as it is to knocks and scrapes. In terms of aesthetics, this is a question that everybody can decide for themselves when they see the accompanying photos.

Step by step

New soles

➤ *Samba's original cabin soles were of quite high quality, with teak strips glued onto a plywood surface. As time had passed these had become unstuck, some had even been lost and then replaced with new strips. In order to restore these we would have to have stripped the teak down to recover its natural tone and then revarnished it. This is hard work when you are using two-part varnishes which, in our case, would have been necessary as the teak had been darkened by the action of the sun's ultraviolet rays. We would also have had to clean up the underlying plywood boards, repaint them to protect the wood from the damp of the bilges, and would also have had to replace some of the cross beams that were broken or had become detached.*

➤ *We finally chose to install marine plywood boards lined with a highly resistant synthetic panel that included an imitation teak strip veneer. These boards are quite expensive but are very easy to install. All you have to do is trace out the shapes of the old sole boards, cut the new sections to size, fit them in place and they are ready to use. Because the underside is also lined with synthetic veneer it does not need to be painted or varnished. All you have to do is apply epoxy resin or varnish to the edges of each section, where the plywood is exposed, to protect it from water and damp. This type of sole has become increasingly more commonplace in production boats, as it offers a number of advantages, such as not needing maintenance, being resistant to most physical and chemical damage and the fact that it will never need to be varnished. What an invention!*

➤ In order to ensure a clean cut, when cutting plywood boards that incorporate a synthetic finish, you have to use special saw blades. With normal wood cutting blades you can easily damage the synthetic finish during the cutting, which will spoil the look of the sole when it is fitted. If in doubt, it's best to make a test cut on one of the areas that is going to be scrapped anyway, before going ahead with the final cutting.

➤ For aesthetic reasons we were obliged to ensure that the lines of the strips were maintained throughout the whole of the cabin space. The easiest way to achieve this was to fix the old sections of sole to the new boards, in the same position that they would occupy when installed on board, then mark out the edges and cut the sections. We recommend that you study the position of the parts, before starting to mark them out, in order to make the most of your boards, as they are far from cheap. Before removing the old soles you should also consider any modifications or adjustments that you might want to make in terms of the size and/or layout of the sections.

➤ Having marked out the sections, cut them using an electric jigsaw, which is more practical for this work than a fixed circular saw would be. Take care when cutting, use sharp new blades and do not force the saw, allowing it to advance at its own pace, carefully following the lines marked out. Piece by piece the sole will start to take shape, a carbon copy of the original.

➤ With the pieces now on board (the previous steps took place in the carpenter's workshop), the time had come to make the last adjustments to the finishes. Do not forget that on a boat, although at first sight they may appear to be so, there is hardly ever such a thing as a straight line or a 90° right angle. For precision adjustments, like those for the soles, you always have to allow for slight adjustments to be made before the item will fit neatly into place, rather like a large jigsaw puzzle.

➤ The edges of the undersides of the sections that rest up against the hull had to be set off at an angle to adapt to the shape of the hull. Our carpenter used an electric planer to do this, although this is a tool that requires quite a lot of practice before it can be safely used by an amateur.

➤ One of the advantages of plywood boards lined with synthetic material is that resistance to bending is much higher than that of plain plywood. On board Samba the base of the existing soles was a 15 mm plywood board plus 3 mm thick teak strips glued to the top surface. The total thickness of the new panels was only 13 mm and they were more resistant to bending than the old ones were. In any case we also attached the longer sections, or those with less support, to the wooden crossbeams using screws and polyurethane adhesive.

➤ *The hidden fittings that can be found on many boats for lifting sections of the cabin sole look great when you see them at the boat show but, in the long run, they just get rusty and break or fill up with rubbish and end up being unusable. A few simple 20 mm diameter holes hidden away in the corners is a practical, long-lasting and a very cheap solution.*

➤ *In just over a day's work we managed to completely renovate Samba's cabin soles. The change could not have been more spectacular and would have looked even better in the photo if we had removed the protective film that covers the surface of the boards. We decided to leave this in place as a precaution until we had finished all the work on board.*

Restoring the saloon table

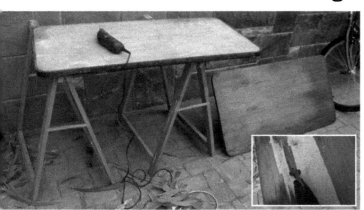

➤ *The biggest problem with the table was the peeling and scratched sapele veneer in some areas and the damp-blackened patches in others. Using an electric scraper in a couple of hours we had removed this old veneer and then, first using stripper and then with different grades of sandpaper, we managed to remove all traces of the old glue and varnish.*

➤ *The most complicated part of cutting this veneer was to make sure that it fitted precisely inside the raised rim moulding. Bit by bit the carpenter worked on the rounded edges of the corners, making sure they were a perfect fit for the raised rim. This wing was only rectangular in the loosest sense of the term. None of the corners were the same as each other and were not cut at perfect right-angles, while the sides were not precisely parallel. Getting the veneer to fit perfectly was the work of a craftsman. Immense care also had to be taken in order to avoid breaking the pieces of veneer, which were very fragile right up until we had them firmly glued to the wooden base.*

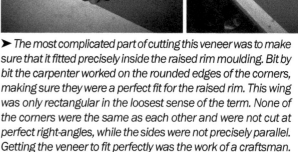

➤ *Rather than replace the fine sapele veneer we chose to use a synthetic finish. These fine linings (0.7 mm thick) are sold by specialised carpentry shops and DIY stores. Basically they can take whatever you throw at them which is just what you need for a saloon table that will be exposed to all manner of scratches, damp, staining and knocks. For the table's two wings we first cut the veneer to the approximate shape, working from a sheet of 1.20 x 2.40 m.*

➤ *When the veneer was fitted into place, and before the glue had time to set, we placed the two wings one on top of the other and left them overnight in a hydraulic press. When working with synthetic veneers you will need a number of specialised tools. Many carpenters, particularly those who specialise in kitchen installations, have the necessary tools for working with marine plywood and synthetic veneers. It is important to mention this as the base of the synthetic prefabricated veneers that are usually found at DIY stores is chipboard, a type of material which does not adapt at all well to a marine environment, where it tends to lose consistency. Also make sure that you use glues that are resistant to humidity.*

➤ With the veneer firmly glued to the two wings of the table, the finishing of the curved edges was achieved quickly using a band sander.

➤ Another specific machine for working with synthetic veneer was this sander that blunts its sharp edges; otherwise you could end up with a potential knife-edge all around your table.

➤ The above photos show the final finishing touches on the table, redoing the varnish of the moulding trim, the edges of the plywood and the undersides of the table. As you can see in the first photo, the contrast of the teak with the veneer is quite pronounced. In order to match them up better we dyed the veneer with a teak tint and finished the whole assembly off with three coats of varnish.

➤ This photo shows the final result of our saloon table restoration. When unfolded it is just the right size for a 40-footer (1.15 x 1.10 m), particularly if you take into account that in the saloon we have five metres of seating space surrounding the table.

Replacing the tongue and groove

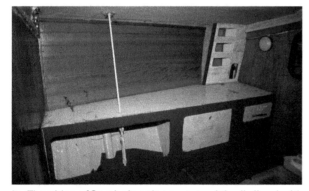

➤ The sides of Samba's saloon were originally lined with pine tongue and groove planking. Attempting to restore this kind of lining is just throwing your money away. It is far easier to completely replace it. You can find this type of planking in any DIY store, with an endless range of colours and finishes at easily affordable prices. The only requirement for a boat is that these should not be too wide (at most 10/15 cm) or they will not adapt to the curves of the hull. It is also important, particularly if you go for the cheaper ranges on sale in your local DIY store, to allow for more planking than you are actually going to need to cover the surface (one and a half times), given that wastage resulting from bothersome knots and other imperfections is likely to be quite high.

➤ The hardest part of the work is the part that you never see. Due to the impossibility of screwing the planking directly to the hull you have to fit wooden chocks or battens at regular intervals (every half metre or so) over the surface to be covered, so that you can screw the planking to them instead. In the case of Samba this work was simplified because there were various reinforcements along the length of her hull that were ideal for this purpose. Where there were no such reinforcements, and in order to bend the battens to the shape of the hull, without them breaking, the carpenter made a number of short crossway cuts. If you do this, the batten will easily bend to adapt to the shape of the curve. You can also achieve the same effect by gluing chocks to the hull using polyurethane adhesive.

➤ Once we had made the bases for screwing on the planking it was simply a question of cutting the planks to size and fitting them into place (planks are usually supplied in 1.90 to 2.40 m lengths) starting from below. In the case of Samba, the inspection of the electrical wiring ducts at the top of the side walls, as shown in the photo, could be done by simply removing the top plank. We also trimmed off the tongue on this last plank, which meant that it could be put in place and removed for this purpose with ease.

➤ This is what the saloon looked like after we had finished lining it with the new tongue and groove planking. The most observant among you may have noticed that we took the line of the ceiling as our horizontal, when it would have been more usual to have taken the sole. This lack of orthodoxy, however, had left the removable top plank, behind which the electric wiring runs on the port side, free for checks. If we had taken the sole as our horizontal then the top plank would have had to be cut at an angle, complicating wiring checks.

➤ We also lined the sides of the fore cabin with tongue and groove planking painted white with a satin finish. Lining the walls of the boat was less than a day's work for the carpenter. This was an economical solution that is also rather attractive.

Chapter 35

Engine (part 2)

Describing a change of engine as restoration would be cheating. The authentic thing to have done would have been to take the old engine to a workshop, strip it down and give it a complete overhaul. But, as we saw several chapters ago, any attempt to save Samba's original engine would have been a pipedream.

Engine replacement

Few amateurs will have the technical skills and know-how required to install a new engine and, bearing in mind that the guarantees offered by the different makes, logically enough, are only binding if your engine has been installed by an officially approved and qualified mechanic, nor should they attempt to do so.

However, being a first-hand witness to the installation of a new engine is an interesting experience. With your boat in the water and no mechanic in sight, a perfectly normal situation on a sailing weekend or during a summer cruise, it is essential that all boat owners should at least know the basics of how the component parts (supply, filters, exhaust, batteries, alternator, cooling, etc.) of their engine work, where they are and what to do when something goes wrong.

Now is not the time for a course in emergency engine repairs but the first lesson regarding the things that an amateur can do (and there are a few), as well as aspects of basic and periodic engine maintenance, involving the identification and location of those parts that may, on occasions, fail.

Engine replacement: loose ends

✓ There is not much point in installing a more powerful engine in a sailing boat with the idea of making her faster. The speed limit, in knots, is always going to be around 10 or 20% below her length at the waterline in metres (although planing hulls can surpass this figure). A yacht such as *Samba*, with a waterline length of 9.16 metres (just over 30 feet), would find it difficult to make more than 7 or 8 knots, regardless of how powerful the engine she carries is.

✓ Each boat has top and bottom power limits that are reasonably adapted to its characteristics. At the bottom end, consumption, weight and price are minimised, while at the top end you may gain a little in speed and power against the wind or the swell.

✓ Whenever you change the engine you will also have to change the propeller for one with the right pitch for your new engine. Professionals in this field work out this calculation on the basis of the boat's characteristics (weight, length, etc.), installed power, the reduction provided by the new engine's gear box and the maximum possible diameter and type of propeller to be installed.

✓ Do not forget to order a propeller that functions with the correct turn direction (right or left). If you get this wrong, something that happens more often than you may imagine, you will go faster in reverse than in forward gear.

✓ In this chapter we shall be concentrating our attention on the installation of the engine itself but, in order to start the engine, we also had to install a renovated electrical system, with specific circuits for the alternator and the batteries (start up and services).

Step by step

➤ For an engine to work without excess noise or vibrations, it has to be perfectly aligned with the prop shaft. The engine beds, stern tube and stern bracket must be accurately aligned. When replacing the old engine, the most tried and tested solution (also the most costly) is to start from zero, rebuilding the beds and stern bracket to adapt them to the new mechanics. However, in view of the good condition of these components, and taking advantage of the slight difference in terms of level between our old Perkins and the new Yanmar, the mechanic chose to redesign the new engine's brackets, so that they would fit onto the old beds.

➤ The mechanic ordered four custom-made brackets for the engine. These pieces, which attach the block of the engine to the silent blocks, were made up to match the separation and height of the existing beds, which meant that we did not have to redesign them.

➤ Previously, engines were often bolted straight onto the beds. When anchored in this way, the engine's vibrations attack and weaken the fibreglass and wood within a relatively short period of time, leaving the engine loose on the beds and only held in place by its own weight. More often than not, on boats of a certain age, this is the explanation of their engine vibration problems. Silent blocks do not have a base to damp their vibrations and thus transmit them to the whole boat. To avoid this problem, we fitted inverted 8 mm thick U-shaped iron brackets, bolted to the beds, to ensure that the engine would be firmly anchored.

➤ When the time arrived to install the engine, the crane lowered our brand-new diesel onto the beds. At first no attempt was made to definitively line it up and then, little by little, as the other parts and fittings were lined up, it was connected and adjusted until final alignment was achieved with the shaft and the stern bracket. When this had been done the engine could finally be firmly bolted in place.

➤ The inverted U-shaped brackets were firmly attached to the beds by means of through-bolts, from one side of the bed to the other. The engine's silent blocks were then bolted onto the threaded hole in the U-shaped brackets. This attachment was so strong that, by taking hold of the engine, you could almost lift the boat.

➤ The next step was to assemble the stern tube; although this part looked new it was actually the original fitting. The new graphite fibres that we had used to pack the tube had to ensure that this part continued to function for a few more years. Before fitting the tightening clamps, to ensure it sealed tight, the mechanic applied a little polyurethane sealant inside the rubber pipe of the stern tube. For sailing boats with a fixed fin on the rudder blade, as is the case with the North Wind 40, before installing the engine you have to fit the prop shaft in place from inside the boat. There is no other way to do this and if you install the engine first you will have to take it out again to fit the prop shaft.

➤ In order to attach the prop shaft firmly to the engine, first the plate was coupled up and then bolted to the engine's gear reducer. This was a perfect moment to check for shaft alignment, as the plate acts as a perfect visual reference. If, when you bolt the plate, you have to force the stern tube or the stern bracket, even only slightly, this is a guarantee of subsequent mechanical vibration and premature wear and tear on the stern tube, the stern bracket, its bearing and the prop shaft itself.

➤ When it came to mounting the instrument panel we took advantage of both the old location at the foot of the cabin entrance stairs and the attractive solid teak frame that had held the instrument panel for Samba's original engine. A thorough stripping, a meticulous sanding down, a piece of plywood mounted behind to support the new panel and four coats of varnish served to recover the control panel's original distinguished air.

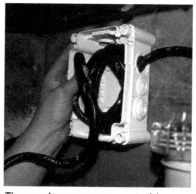

➤ From the back of the instrument panel a bundle of wires emerged, through which information is sent by the engine to the panel (start up, rpm, temperatures, oil pressure, shutdown, etc.). These wires were connected by means of special jacks. In our case the panel was so close to the engine that we ended up with a lot of spare wire; the engine and control panel were delivered with a length of around 4 metres, allowing the panel to be installed in the cockpit if necessary, something that is fairly common in many modern boats. Instead of leaving the wires hanging loose or coiled up in the engine room, we hid them away in this connection box: a cheap and discreet solution.

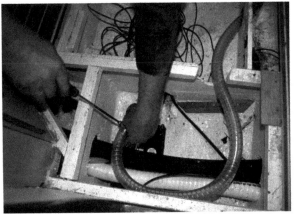

➤ The next step was to connect the cooling circuit, through which seawater circulates through a filter and round the engine. The cooling water filter has become an almost essential accessory in modern boats. Years ago, a simple grill over the water inlet was all that you needed, but these days a complete barrier of filters is needed to prevent the entry of the kind of plastic waste and detritus that unfortunately you will find just about everywhere. If you do not take care, this will clog up your engine's cooling pump. For extra safety we double-clamped the through-hull hose connection.

➤ The remote control, as with many of the engine's other peripheral components, was installed taking advantage of its original location, in this case on one side of the cockpit next to the helmsman's bench. Good access to the mechanism of the push/pull control cables was provided by means of a trapdoor located in the stern cabin.

➤ In most boats the installation of a drainage valve is recommended and for V-shaped hulls, where the engine is located well below the waterline, these are essential. The drain valve, due to its siphon effect, prevents possible flooding of the boat, due to the entry of water through the cooling circuit, when the engine is switched off. This should be installed as high up as possible in the engine room.

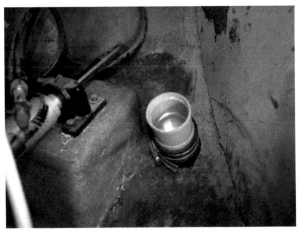

➤ Although we completely renovated the component parts, for the exhaust we also took full advantage of the original pipe and silencer emplacements. The photo shows the exhaust outlet from inside the boat, with the connection to the exterior plate ready to receive the final section of pipe.

➤ We also fitted a siphon to the exhaust outlet (impossible to get a good photo of this) up under the helmsman's bench. This also serves to avoid water stagnating in the circuit. The end of the siphon was connected to the silencer and, from there, the hose runs to the water manifold, alongside the engine.

➤ We could not find a water manifold that was adaptable to the engine casing. Some models were too big, others too wide and others too high. Fortunately almost every problem has a solution and, in the end, the mechanic arranged for this custom-made manifold to be made up in stainless steel.

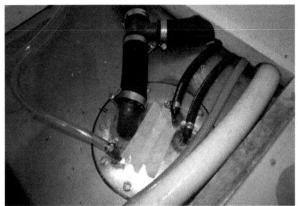

➤ With the exhaust set up completed, it was time to think about fuel supply. Samba is fitted with two stainless steel diesel fuel tanks. The above photo shows the filling hose, which has an overflow bypass and a T-fitting leading off to the other tank. This system, although not much used, does have its advantages, avoiding the fuel sloshing about between one tank and another when sailing, as tends to be the case when the two tanks are joined at the bottom by communicating hoses, and there is no shutoff valve in the bilges. The photo also shows the fuel feed lines and the return and breather hoses. In the central opening, covered here with tape, we will install the flow meters, once restored.

➤ The double tanks with top-mounted fuel intake require a double supply intake but, to compensate, allow for the engine to be supplied without any problems from either of the two tanks, even when heeled over hard. In order to be able to alternate the flow of diesel in a simple and effective fashion, we installed the two-way valve you can see in the photo. A single movement of the lever will change the supply to the tank you require.

➤ The fuel line passes through the decanter filter which, as well as pre-filtering the diesel fuel, also separates out any water that may have found it's way in when the tanks were filled, or have formed as a result of condensation once inside the tanks. From this filter the supply line passes directly to the engine.

➤ Having mounted all of the peripherals, the installation could be taken as practically finished. An important last step, the absence of which has led to the lamentable ruin of many a new motor, is to fill the engine and inverter with oil (they are delivered empty from the factory) and also fill up with cooling liquid.

➤ This is what the engine room looked like with the engine all set up and ready to go. For the installation of the peripherals we ensured that those elements that require periodic checks, particularly water and fuel filters, were easily accessible.

➤ We ended the installation by fitting the propeller, a model with two fixed blades. We would have preferred to mount a folding or variable pitch propeller but neither of these were possible, as the limited distance between the end of the shaft and the rudder skeg only left room for a fixed propeller. At some point in the future we will be taking a closer look at trying to resolve this problem.

Engine replacement: balance sheet for the operation

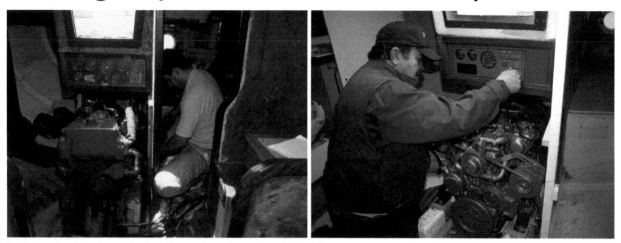

➤ *Of our budget for replacing the engine, the engine by itself accounted for the greater part of the cost (65%). Peripheral accessories (filters, exhaust system, shaft, plates, etc.) also cost quite a bit (25%) while, unusually, the costs for skilled labour represented a very low percentage (a mere 10%) of the whole operation. From the moment we started to disconnect the old Perkins until starting up the new Yanmar, we invested 6 or 7 days of work. As a result of this change we gained 15 HP of power and, paradoxically, reduced the size of the engine block in terms of both length and height, gaining space for the saloon. This is one of the marvels of modern technology.*

Chapter 36

Lining ceilings and walls

As time goes by, ceiling and wall linings, which were all the rage in sailing boats a few decades ago, tend to deteriorate and come unstuck. Replacing these linings is a job for either professional tradesmen or extremely gifted amateurs.

Plenty of craftsmanship

One of the reasons for the gradual disappearance of upholstered linings in production boats is the high cost of both the materials and the skilled labour required to fit them. Vacuum or infusion lamination techniques have created a new generation of boats with interior finishes that are so smooth that they do not require any decorative linings. These days, with just a couple of carefully located mouldings, you can easily achieve an interior aspect that a few years ago would have taken hours and hours of work; a healthy saving for both the shipyard and the buyer.

This was not the case when hulls and decks were laminated using only female moulds, their interior surfaces having the coarse finish of exposed fibreglass. Concealing this finish in bilges, the bottoms of lockers and backs of cupboards can be done rapidly and cheaply with a couple of coats of paint. But the so-called 'prominent areas' require more elaborate treatment. The two most frequent solutions are panels, attached to the hull with screws, Velcro, etc., or glued-on textile linings.

Any solution, as long as it is well installed, can provide good results and each one has its advantages and disadvantages. If you are thinking of restoring a boat of a certain age, because radical changes or adaptations usually turn out to be such a pain in the neck, the simplest solution is usually to stick with the original system, which was what we did when it came to restoring *Samba*'s ceilings and walls.

Samba had been fitted out with a number of panels (some original and some installed at a later date) lined with synthetic material, on the saloon ceiling, plus a carpeted liner that had been directly applied to the ceilings of the cabins and the head. The only problem with the ceiling panels in the saloon was their age, but carpeted linings are an example of a material that ought to be avoided on board.

Carpeted linings, even those that theoretically are specially prepared for marine environments, absorb water, humidity and dust. They almost always end up giving off a characteristic and rather unpleasant odour that, little by little, ends up impregnating the whole interior. Despite their attractive price, it is better to avoid them, particularly in view of the fact that a complete range of synthetic materials is available, which are truly adapted to on-board conditions, with medium-term advantages that far outweigh the difference in cost.

To line our ceilings we chose a 100% vinyl fabric. This lining, which can be used either below or above deck (ultraviolet protection), is rot-resistant, extremely hard wearing (12 times more than normal PVCs), incorporates a fire-retardant treatment, is sound-proofed and easy to clean, even using solvents.

The lining we chose had a slightly cushioned polyester fabric base, but for the lining of the walls and ceilings we ordered the necessary quantity padded with a fine (4 mm) layer of foam. We did this in order to increase the thermal and sound insulation and, at the same time, better disguise the irregularities in the interior surface of the hull. In total we used 7 metres of normal fabric and 20 metres of the cushioned fabric. As a rule of thumb, to line the interior of a boat you need an area of material roughly twice its length.

At first glance installing the interior lining of a boat does not look that complicated, it's just a question of cutting out the pieces and gluing them to walls and ceilings with contact adhesive. The problems start to arise when you have to cover pronounced curves, joining different pieces of lining, edge work where pieces come up against bulwarks or furnishings and the necessary seams and finishes required for a lining that is going to be permanently visible. With this kind of work there is no room for improvisation and only experienced professionals will acquit themselves well.

The work that you will see in this chapter took about a week to do. What with the raw materials and the labour costs, lining the interior of a boat is far from cheap. Before deciding on a specific fabric or upholsterer it is usually a good idea to see the material in place or check out previous jobs that your professional has done. If you can have a look at a boat that he has lined using the same material you will have a very accurate idea of what that combination is going to achieve on board your boat.

Step by step

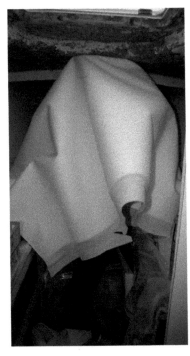

➤ *After we had cut out the piece of material to the approximate dimensions of the area to be lined, in this case almost half the ceiling of the forward cabin, we started to glue it in place. To do this correctly you have to start in the centre of the space to be covered. Amateurs are always tempted to start on one side and work across to the other. This is a bad idea. If you do it this way you are almost guaranteed to end up with it hopelessly misaligned.*

➤ Once the piece has been glued to a spot in the centre of the area to be covered, we then started to glue it in place, little by little, working towards each of the sides along the most elongated axis of the piece, until we had finished. The vinyl fabric that we used was quite elastic, which means that it adapted well to the small curves and irregularities without creasing.

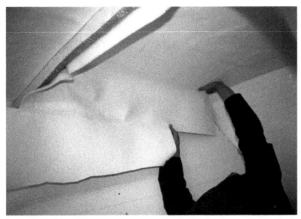

➤ Starting in the centre and working towards the sides, little by little the upholsterer completed the textile lining of the forward cabin ceiling.

➤ To line a pronounced curve on the roof, without letting the material sag, you first have to glue it to the irregular area. As soon as the glue has set and offers greater adherence (approx. 15/20 mins), continue working the piece to the end. As you only apply the glue to the roof, you hardly need to protect the furnishings, although this is a necessary precaution when you apply the glue directly to the fabric, with the consequent risk of leaving smudges of glue all over the place. When working with contact glue it is important to keep the boat well ventilated as the fumes can cause headaches and dizziness.

➤ Excess fabric that is left round the edges can be tucked under the furnishings using a putty-knife or trimmed off with a cutter or scissors. In some cases you may decide to use a wooden strip to cover the edges.

➤ A coat of glue on the ceiling or hull, previously cleared of any traces of the original glue or lining, is sufficient to hold the limited weight of the textile lining. In those areas where greater adherence is required, glue can also be applied to the fabric. A good trick for avoiding smudges on the furnishings is to first of all glue the lining just to the ceiling and then, before the glue dries, partially or wholly pull the piece away. The marks left by the glue will indicate where you need to add a supplementary coat of glue to the fabric.

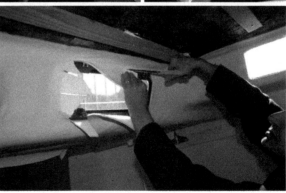

➤ As a general rule, when lining areas of the boat that include companion ways, trap-doors or hatches, do not make your final cuts until the last minute. Trying to measure out and pre-cut the opening accurately is virtually impossible. The best thing to do is, little by little, make radial cuts, from the middle of the opening towards the edges, as you attach the lining to the wall or ceiling. When you finally have the piece glued in place, you can finish by cutting out the outline of the opening using a cutter or scissors, avoiding the risk of making mistakes and spoiling the whole piece. In cases where the liner has to go underneath light fittings, frames or any other elements that are screwed into place, one trick for finding the screw holes again is, when removing the part, put the screws back in their holes. Later on it will be easier to find them by touch.

➤ On the basis of what we have seen up to now, the interior lining work still does not seem that complicated. Do not kid yourself. If the elasticity of the fabric is sufficient to disguise the curves and corners of the hull's small irregularities, sooner or later you are going to come across a point where seams will be required, two pieces of fabric will have to be joined, or a combination of both, in order to elegantly resolve the point where two lined sections come together. When that happens, only the skill and know-how of an experienced professional is going to ensure an impeccable finish. In the photo, for example, we can see where the side and central pieces in the saloon came together, with the need for a seam in the inside corner of the ceiling, another on the frontal finish and a fold that has to be hidden along the length of the bulwark reinforcement strut. The kind of skill and experience needed to overcome such geometrical complexity is not within the grasp of your average DIYer.

Upholstering the stern cabin

➤ Stern cabins, which are pretty much the same in most boats of Samba's size, are very difficult to line. For the best results, from an aesthetic point of view, we started off by sewing the piece that will make up the forward bulkhead, which then had to be extended along one side of the ceiling, in this way providing a good finish for the most visible part of the join. This piece was later glued to the cabin ceiling, hiding the edge along the top part of the cockpit coaming, the part of the berth that is least visible.

➤ The next step was one of the most complicated; to sew two pieces of upholstery together in the shape of the underside of the cockpit benches and which, in one piece, would line the rest of the ceiling and the side of the cabin. You do not have to be a DIY wiz to get an idea of how difficult that was to do. After this, all that was left was the end of the cabin, the only flat surface in the whole of this space.

Finishes: a question of detail

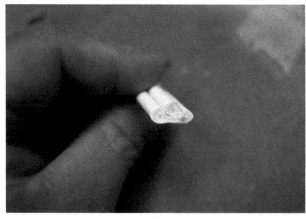 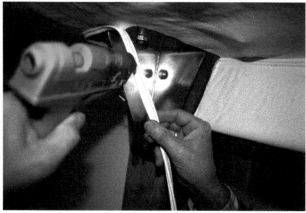

➤ Once we had the interior fully lined it was time to start working on the finishes. To cover over the edges where the pieces of vinyl lining meet the hull, the upholsterer made up lengths of trim by sewing together three fine cords wrapped in a strip of the lining fabric. The advantage of this triple trim is that it does not crease, no matter how much you bend it. The upholsterer then glued this trim in place between the hull and the fabric, achieving an attractive overall finish.

➤ To glue the trim in place you can use either hot instant glue or polyurethane adhesive. The latter is a bit more awkward to apply, but much more resistant in the long term.

➤ Another solution that we used, in this case to disguise the edges along the false bulwarks, is what in Spain we call 'a la inglesa', or English style. This system involves first of all gluing a strip of vinyl onto a fine (4 mm), narrow (1 or 2 mm) plywood strip. This strip is then turned on itself and attached, using small steel tacks, to the bulkhead in such a way that it completely covers the edge of the piece of ceiling lining. Later, giving it a turn on itself, the lining ends up perfectly adjusted to the ceiling.

A spectacular change

➤ The photos above illustrate, better than words ever could, the vast improvement that resulted from lining interiors. As with many other areas of our refit, the quality of the work and the attention to detail are as important, if not more so, than the material or the colours chosen. Once we had completed the interior lining we could finally fit the interior trim round the portlights, which now serve to cover the edges of the lining.

Lining the ceiling panels

➤ If the lining glued to the walls and ceilings required professional know-how and skills, the re-upholstering of the ceiling panels on just about any boat could be managed by any reasonably competent amateur. When we removed the old panels we replaced the chocks from which they hung with battens. These are more versatile when it comes to aligning the exterior screws.

➤ At the entrance, the sheets of plywood that we attached to the ceiling, as reinforcement for the deck hardware, lowered the height of the ceiling by about 3 cm. To recover some of this height the upholsterer cut out holes in the relevant panels, over the ends of the bolts and the nuts. This allowed us to recover 1 cm. The rest, as you will see in a later chapter, was left to the carpenter.

➤ The upholstery of the ceiling panels is not glued onto the visible side, which means that, when the panels are in place, the fabric will hang a little loose, free of the panel, giving it a more cushioned look and avoiding marking out any imperfections on the plywood panels. To glue the fabric in place apply a 5 cm band of glue all around both the top edge of the panel and the fabric. After 10/20 minutes, when the glue is no longer tacky to the touch, the lining is pressed firmly into place. In order to centre the piece and avoid the creases created when you pull the fabric more towards one side than the other, instead of starting to glue at one corner and follow the edge of the panel, fix the fabric first by tensioning it slightly in the middle of each of the four sides of the panel, and then work from these points towards the corners.

➤ For the finishing of any recessed corners make a cut, almost as far as the wood at the bisector of the angle. The fabric is subsequently folded over and glued to each side, stretched slightly at the vertex of the angle. When seen from below, the finish looks perfect.

➤ For the exterior corners first of all two cuts are made in the fabric, but again not as far as the wood. This will allow the two sides of the corner to be glued, leaving a rhombus-shaped piece. Pulling on the tip of this piece it is also glued to the panel. All that is left to do is trim off the excess fabric until all that remains is a small strip.

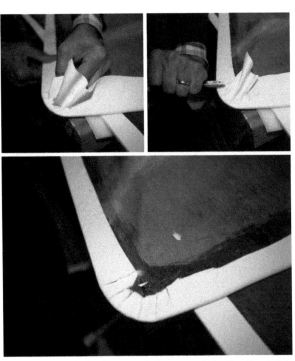

➤ The best solution for rounded corners is to glue the fabric down, bit by bit, making pleats around the rounded part, regularly spaced out round the radius and taking advantage of the elasticity of the vinyl liner to establish the uniformity of the curve when seen from below. Once the fabric has been glued all that you then have to do is trim off the excess, so that it does not form a thicker wad at the corner that would make it more difficult to fit the panel neatly in place.

Fitting the panels

➤ Before fitting the panels in place again, use a piece of tape to mark out the position of the supporting battens. If you do not do this, the panels will be hidden from view and it will be impossible to see where the screws have to be fitted. The first panel to be fitted is the central panel in each area. Once this has been provisionally fixed in place with tacks, the process is continued towards the sides of the boat.

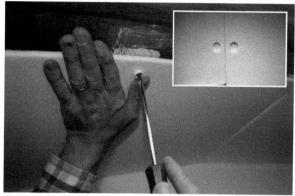

➤ Place the tacks symmetrically and/or in line. When all the panels have been fitted into place all you have to do is replace the tacks with the final screws. We covered the screw-heads with white plastic caps to hide then from view and, at the same time, to protect the upholstery from possible tears. An equivalent effect can be achieved using stainless steel tooth washers.

➤ The final result of the lining was visually quite stunning. The white colour increased the sense of spaciousness, light and ceiling height. In the case of Samba, this left nothing to be desired.

Chapter 37

Anchor locker (part 2)

We finished the renovation of the anchor locker by making up a new cover, installing the windlass and replacing the mooring cleats.

Readying the bow

In the first chapter on the anchor locker (Ch. 10) we took a look at the most laborious part of the renovation, laminating the support that the windlass was going to be bolted on to. We also solved the problems of the rigidity of the frame but left pending the replacement of the covers, both wings of which had suffered knocks and delamination, the installation of the windlass and the replacement of the mooring cleats.

Instead of renovating the original double-wing cover we chose to make up a new one-piece cover with a hinged opening to one side: a more modern

and functional system. We also installed new mooring cleats. *Samba*'s original large centrally-mounted cleat was as strong as an ox but pretty inconvenient to use. The new layout is standard for most sailing boats, with a cleat on either side. Once this work had been done the bow was ready for putting to sea.

When you are refitting or restoring a boat, it is inappropriate to use the word 'finished' to refer to the conclusion of any job. A boat, by definition, can never be considered finished, she is simply ready to sail. There's always something that needs to be done and, if you do not know what that is yet, you will soon discover it. Every owner has, either in mind or marked down in a notebook somewhere, a list of jobs that could be done or have to be done on his boat. Our list was a long one

and was going to be even longer as soon as we got her back on the water. At this point, let's just say that the anchor locker was now ready to go to sea.

Step by step

➤ In Chapter 10 we enlarged the opening to the anchor locker and installed a support for the windlass. We also removed the enormous central cleat which, despite being tremendously solid, was in fact rather inconvenient to use.

➤ The laminator went back to work on the bow by measuring up the outline of the new anchor locker cover. First he made up a template from a piece of wooden sheet and then used the template to trace out the shape of the cover on a piece of 13 mm plywood. This was then cut out and taken to the boat to make sure it fitted, ensuring there was a gap all around the perimeter to allow for the different layers of fibreglass that would later be laminated onto it.

➤ One of the more subtle aspects of this particular job was to recreate the almost imperceptible curve of the deck (common to most sailing boats). With a strip laid across the deck, as shown in the photo, the laminator then marked the line of this curve with a pencil.

➤ This curve was then reproduced on four crossbeams, which were attached, with glue and screws, to the underside of the cover, forcing it to bend to form the same curve as the deck. These crossbeams also served to strengthen the cover.

➤ With the slightly curved crossbeams in place we again checked the fit of the cover and defined the line of the anchor chain at the front end of the cover, marking out the form with a block of rigid foam. The foam could then easily be cut and shaped to the required dimensions. Once its form had been moulded it was glued to the cover.

➤ Having finalised the design, our laminator started to laminate the cover, applying five successive coats of fibre and mat to both sides. The anchor locker is one of the parts of a boat that is most exposed to the spray and you cannot take too many precautions when it comes to trying to keep it dry.

➤ Before starting work on the lamination we attached strips of wood all around the edges of the cover, reinforcing it and improving its fit in the frame of the locker. Once the first coats of fibreglass had set on the topside of the cover the time had come to cut out and shape the opening for the chain, also shaping the interior notch through which the chain would run.

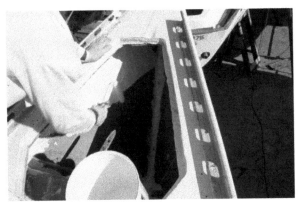

➤ In the meantime we also took advantage of the opportunity to give the interior of the anchor locker a couple of coats of paint. The shine that a couple of coats of paint will give to the bottom of lockers and bilges never ceases to be a surprise. The white colour highlights and protects all the work that has previously been done with fibreglass and laminates.

➤ With the anchor locker painted, and the cover finished, this was the perfect moment to install the anchor windlass. In mechanical terms this was relatively simple. Once we had made the holes for the through-bolts, the motor and the passing of the chain (template attached), and after sealing the upper housing, to prevent water getting to the motor, the capstan was finally fitted into place. All that was then required was for the bolts attaching the motor from below to be tightened to the best of the five possible positions.

➤ The windlass' electrical connection was perfectly illustrated in the installation manual. It was a question of connecting the thick 35 mm cables from the battery, via the cut-out, to the relay box, switches and windlass motor. Instead of fitting the relay box in a corner of the locker we installed it in the interior of the boat, set off in one corner of the forward cabin, safe from water and humidity. Water getting into relay boxes is one of the main causes of windlass malfunction. On some boats, the windlass motors are even isolated from the anchor locker and only accessible from the forward cabin. This is even better!

➤ The locker cover, once it had been laminated, was given two coats of epoxy resin. After filling and sanding down the top of the cover we applied two coats of polyurethane paint followed by a final coat of non-slip paint.

➤ With the cover in place we fitted the latch and hinges. The latch was a stainless steel fitting with a rotating catch. To install it we made a hole, the position of which was defined by the length of the catch that fits under the lip of the locker. This distance had to be calculated earlier. We used the hinges from the original locker covers. Although this may often appear not to be case during Samba's refit, we actually took advantage of the original hardware and fittings as often as possible, even knowing, as was the case with the hinges, that buying new ones would hardly have cost us anything.

➤ After we had sanded the top edges of the hole down a little, to adapt them to the bevelled edge of the latch, we fitted it into place with four screws. We also added a little polyurethane sealant in order to guarantee watertightness, cleaning off the excess before it had time to dry. We also applied sealant around the inside of the latch hole, to protect the exposed wood of the cover from water.

➤ Where the adjustable catch of the latch made contact with the frame we also used polyurethane adhesive to glue on a small wedge-shaped block. This avoided the cam damaging the fibreglass as a result of the constant opening and closing.

➤ Our refit of the anchor locker culminated with the installation of two mooring cleats. If you remember, when we installed the support for the windlass we reinforced this area with two pieces of 15 mm plywood. But the core of the deck structure in this area was balsa wood. Balsa wood is a light wood but one that is not very consistent in terms of compression. There are various solutions to this problem, which often arises during this kind of refit. One of these is to use stainless steel or epoxy resin bushings to avoid compressing the balsa wood. We chose a different solution, also effective but simpler, extending the tightening surface using large stainless steel washers between the cleats and the deck. This solution was also aesthetically pleasing.

➤ The balance could not have been better. From the original anchor locker with its damaged covers, the closing tabs delaminated, water leaking in and only a small opening, we now had a modern and functional anchor locker. The new electric anchor windlass was another improvement that Samba now enjoyed.

Chapter 38

Deck hardware (part 4)

Getting the deck ready to put to sea brought to an end the first stage of Samba's complete refit. Ensuring that all the new hardware and rigging would fit had been a long and laborious task but by the end, from the outside at least, she looked thirty years younger.

Objective accomplished

I t is difficult to believe how much hardware you can accumulate on your deck. You have no idea until the time comes to reinstall it all, as was our case after removing everything before painting the deck. In this chapter we will be taking a look at the final adjustments, assembly and detailed finishing work required to make sure that everything was shipshape and Bristol-fashion. You are always going to find details that need refining, improving or

reconsidering, but the work that we had done so far on Samba's rigging and hardware was sufficient for it to do its job, which was none other than to allow us to actually sail her.

As with most of the work that you have seen up to now concerning the installation of the deck hardware, the work in this chapter is also well within the reach of most enthusiasts. There are no great technological mysteries and no need for precision tools. All you need is a little experience, a sound working method and free time at the weekends to be able to manage one or more of the following jobs.

When we had finished the work illustrated in this chapter, following many months in dry dock, we could finally return *Samba* to the water, where we continued the work on her interior. However, it is only in the water that we were able to carry out a series of tests and adjustments associated with her rigging, engine and electrics.

Step by step

A private launch

After many months in dry dock and with the refit on her deck and hull complete, *Samba* was finally back in the water. We sweetened the moment with a glass of cava and a few canapés, shared with the staff of the shipyard who had taken an interest in her refit, which went far beyond the merely professional. Following the launch, which was required so that she could undergo a number of sea trials, all that remained was to finish off the interior refit.

➤ *One of the most symbolic aspects of this preliminary baptism was returning Samba's name to her sides. We ordered the letters of the name, along with her registration, in self-adhesive vinyl. The resistance of this material to salt residue and the action of the sun's ultraviolet rays is guaranteed for many years.*

➤ *Before lowering her into the water, while she was still suspended from the straps of the travelift, was the perfect moment to give the base of her keel, and other areas previously inaccessible because of the struts, a final coat of antifouling paint.*

➤ *Watching* Samba *as they moved her from dry dock and lowered her into the water was an emotional moment for everyone who had taken part in her refit.*

Renewing the deck grab rails

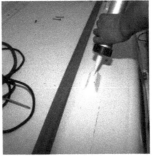

➤ Before starting with the screws we applied sealant to all of grab-rail attachment points. We had to do this first because, after we had tightened the first screw, hardly any space was left underneath the rail.

➤ After cutting the teak rails to the correct length, the carpenter started by adjusting their ends. The aft ends needed to be cut with a bevelled edge, so that they would sit flush with the drip rail, while the fore end had to be sanded down so that they would not snag on ropes or sails and to avoid tripping up the crew. If the grab rails ran straight you could almost insert the screws all at once. In the case of Samba, however, as in many boats, the deck starts to close down towards midships. To start with the carpenter just made the first hole, at the end of the rail, leaving the rest until later.

➤ Gradually and firmly tightening the screws, one by one, we bent the grab rail into shape. The carpenter drilled the holes and added the self-threading screws as the grab rail was lined up. This job took two people (the second does not appear in the photos). Someone has to hold the rail down while the first tightens the screws.

Installing the rigid vang

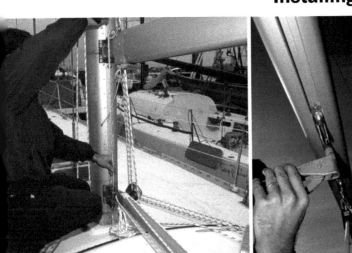

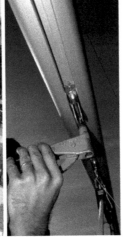

➤ Rigid vangs are a great invention as they keep the boom in place when hauling the mainsail up or down and, above all, when reefing. In our case the boom was 80 cm above the vang mast bracket, which meant that the vang boom bracket had to be located at 120 cm from the mast. All you need to fit bracket in place is a couple of rivets.

➤ Vangs are supplied for the maximum boom length they can handle and almost always have to be cut to size. First of all we raised the boom with the topping lift, until it was just a little past the horizontal, as the weight of the mainsail pulls it down a little. To this measure we added a slight margin, 5 to 10 cm, to compensate for the initial flexion of the internal spring. There is no set size for this cut and, if in doubt, it is better to cut long rather than short – you can always trim more off later – until you find the correct position. The important thing is not to leave the vang too long. If you do, it will stop before the mainsail leech is completely taut, or, if too short, the boom will sit too low.

➤ Once you have the size, you can strip down the vang by simply removing a single screw. The internal mechanism consists of two concentric tubes that compress a stainless steel spring by means of a set of sheaves on the exterior. When cutting off the excess tube you have to make sure that the cut is as straight as possible. Marking the tube with tape is a good way to avoid losing your line when you start to saw through it.

➤ Once the vang has been cut to size and reassembled, all you have to do then is attach the exterior ropes, which will adjust the vang through a system of sheaves.

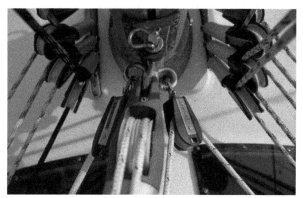

➤ To anchor the vang and mainsail sheets we bolted two eyebolts together, later secured with a couple of spot-welds. To compensate for the overstress on the rivets at the base of the vang we added a through-bolt to attach it firmly to the mast.

New spinnaker pole car stoppers

➤ The oval shape of the mast means that it is inconvenient to install wide stoppers with clips for adjusting the spinnaker pole car. As a result on a lot of boats they use friction cleats, an effective enough system, although one in which the weight of the rope itself tends to work it loose. We decided to install small stoppers that would stop the rope from passing through. It turned out to be quite complicated to rivet these to the mast but, because they do not have to support heavy loads, we decided to use self-threading screws.

➤ Firstly we drilled the first hole, fitting the part provisionally in place and marking the position of the second hole, which we then drilled. After once more checking the alignment of the stopper, we applied a little polyurethane sealant to make sure that the stainless steel screws would not eventually loosen with the vibrations, while at the same time isolating them from the aluminium of the mast. Once this was done we firmly attached the stopper in its place. Before the sealant could set we cleaned off the excess using a damp paper towel.

Mast base halyards

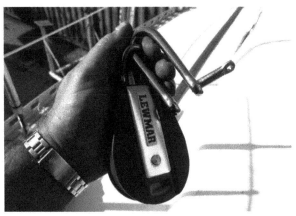

➤ A specific element that we considered was how to overcome a change in level in the run of the halyards from the base of the mast to the cabin roof. To do this we introduced extra-long shackles to the mast foot blocks. The problem was that, simply by fitting them, the blocks tend to slump when they are not under tension, with the risk of kinks forming in the halyards.

➤ When these blocks are attached directly to the deck, the attachments are spring-loaded so that they will be kept tensioned. In the case of Samba, however, the blocks were anchored to an L-shaped plate and it was impossible to install springs. The easiest (and also the cheapest) solution was to fit lengths of hose over the shackles as a kind of support. In this way there was no risk that the block would drop and snag the run of the halyard.

➤ To avoid the halyards rubbing against the paintwork of the deck we also fitted this stainless steel rub-plate guard. These are manufactured in various sizes and serve for both halyards and mooring ropes, or any other ropes that might damage the gelcoat.

A new boom

➤ Depending on the surface area of your mainsail and where it is anchored (cockpit or cabin roof) manufacturers will recommend the type of boom profile most suited to your needs. These profiles are supplied in the maximum accepted lengths. Again, this means that you almost always have to cut them to the required size for your boat.

➤ The next step was to fit the end stops, previously passing the guides for the reef ropes and sail foot. You have to make sure that none of these get crossed over inside the boom. Once the end stops are in place it will be much more difficult to solve this problem. The boom was now ready and its end stops had been riveted into place. For the riveting we used Monel rivets, based on a material that is almost as hard as steel but does not lead to corrosion when in contact with aluminium. Monel rivets are expensive (three or four times the cost of stainless steel) and not easy to find in hardware stores.

➤ All that we had to do to fit the new boom to the boat was to secure the bolt that attaches it to the mast. Remember that we had already installed the bracket, with four rivets on each side, when we were renovating the mast on dry land.

Restoring the spinnaker pole car end stops

➤ The new tab that we welded to one of the car end stops had no holes for bolting it to the mast. One trick for lining up the new holes in the tab is to wrap adhesive tape around the mast, but back to front (with the adhesive facing outwards). Then fit the part in place using the bolts on the other tab and reopen the holes through the tape on the other side. When you have done this, cut the tape and remove the end stop, with the tape stuck onto the mast side. This will clearly mark out the position of the holes. This system will also work for similar cases where you have to line up holes in mouldings or other pieces when working on the interior furnishings.

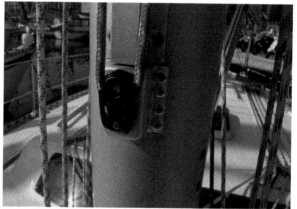

➤ The old spinnaker pole car end stops were utterly unusable. Electrolysis had completely seized their aluminium sheaves while the attachment tabs gave up the ghost when they were being stripped down. At first sight we thought it would be simple enough to replace these stops with new ones, but it turned out that the production of these parts had ceased years ago and finding equivalent parts (height, width, etc.) and adapting them to the spinnaker pole car was so complicated that, in the end, we chose to restore them. Firstly we used the grinder to eliminate all traces of the old sheave and then we welded a new anchoring lug in place and anodised them. Finally we fitted new synthetic ball bearing blocks.

➤ The restored spinnaker pole car end stops are once again in place on the mast.

Self-threading screws

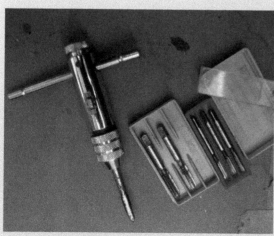

➤ Using metric self-threading screws needs maximum hole-drilling precision with no margin for error. The male parts are sold in sets of three pieces for each thread diameter. The thicknesses are indicated on the boxes along with a second figure, which is the diameter to be subtracted when you make the hole. An M4 mill, for example, is generally accompanied by the figure 70, which means that the base hole should be 40 less 7 mm, i.e. 33 mm. Once the drill hole has been made, carefully insert the male part (they are usually numbered or marked) with the lowest thread thickness. Then insert the following one until the hole has been threaded. Self-threading screws can be used to attach items that are not subject to heavy loads, as the force of the screw will only affect the threaded part which, in an aluminium mast, is not very wide.

Rivets

 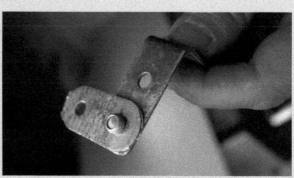

➤ A lot of mast hardware has to be fitted using pop rivets. This is a cheap and easy installation system. All you need are special pop rivet pliers, which you can find at any hardware store. The pliers pull the pin through the rivet distorting it and making the rivet body flare out. The pulling force is resisted and, at a set force, the pin snaps, or pops, leaving the rivet jammed firmly in place. For nautical purposes stainless steel or aluminium rivets are preferred. The former are stronger but, if they are not isolated, electrolysis will form when in contact with aluminium. To fit a rivet you first have to drill a hole in the mast and/or the item to be fitted with the diameter of the body of the rivet. The length of the rivet (only taking into account the body) must be the length of the parts to be joined plus 1 to 1.5 times the diameter of the rivet. With a longer rivet, the body will snap off before the parts have been fixed. If you use a shorter rivet, the ball at the blind end will not press correctly.

Mainsail traveller hardware

➤ Our traveller and car system allows for the modulating of the travellers and their adjustment. For the moment we installed a 4:1 demultiplication for the main traveller and a 5:1 system for the sheet of this sail. When the main traveller track is fitted onto the cabin roof the use of various single blocks is recommended, instead of violin or multiple blocks. In this way the load is shared out on the boom and one-off stresses that could bend it out of shape will have been avoided.

Genoa traveller

 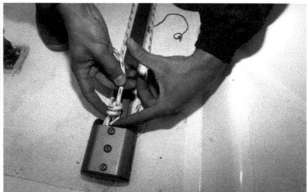

➤ As the genoa traveller track reaches as far as the cockpit, we installed a car adjustment stopper to the end stop of the track itself, with a pull that could eventually be fed to one of the cockpit winches. After inserting the traveller onto the track, we attached the end stop and tied off the control rope. For this attachment the most elegant knot is a running hitch, which will hide the end once it has been pulled tight.

Other details

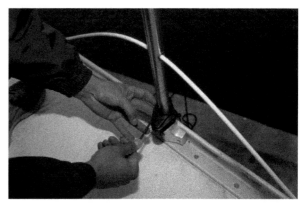

➤ The deck rigging would not be complete without passing the ropes through the reefs. The first reef and the downhaul, the ones used most frequently, must stay on the central sheaves. The furler rope pulleys also have to be fitted into place. We used a synthetic rope which was attached directly to the stanchions.

Installing the forestay

➤ The first part of the installation of the forestay was done with the mast unstepped and on dry land, although the process would have been exactly the same (although more inconvenient) if the mast had been stepped. The mast top attachment for the stay was set barely half a metre from the genoa halyards outlet, with the idea of taking advantage of the genoa halyards and not having to install a specific pair for the backstay. The technician indicated the position of the T-fitting, keeping an eye on frontal symmetry and marking the position of the screws.

➤ When the bolt holes had been made, using a drill bit, a central support hole was also made. The installation was completed by fitting self-threading bolts. To avoid the part falling into the mast, and also as an aid to lining it up, a useful trick is to tie it off with a piece of string that you only remove once the screws have been tightened.

➤ On deck, for the installation of the forestay bracket you first have to find (or possibly prepare) an area that will be capable of handling loads imparted by this stay. A lot of boats, including Samba, have a solid anchoring point over the bow/anchor locker bulkhead. As a fairlead we chose a concealable double ring system, one of which would serve for the turnbuckle while the other would be free to tie off the sail. As reinforcement we also installed a made-to-measure interior base-plate.

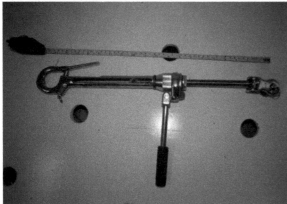

➤ A simple way of avoiding ending up short or long when measuring the length of the forestay (or any other stay for that matter) is to measure the total length of the rope from the point where it is attached to the top to the centre of the bottom pin. Then measure the turnbuckle from the centre of its pin and two-thirds extended, subtracting this figure from the previous one. The forestay turnbuckle that we fitted has a ratchet system that allows for it to be adjusted even when the sail is up.

➤ The original back-stay turnbuckle was a model dating back almost thirty years. Although its retro aspect was quite charming, if you wanted to operate it you needed to have spent a lot of time in the gym. As a result we decided to replace it with a turnbuckle with a crank that you see standing next to it. To install it all you need to do is cut the backstay and pass the two attachment pins. We did, however, avoid the temptation to do it there and then, deciding instead to wait until we had done a bit of sailing, to at least get a first idea of what bow/stern mast adjustments might be required.

➤ As you can see by now, the deck, with all its modern hardware, all shipshape and Bristol-fashion, perfectly disguised Samba's almost thirty years of age.

Chapter 39

This chapter, dealing with the restoration of the galley, is the only one in the book that covers a whole job in one particular area of the boat from start to finish.

Galley: from start to finish

Up to now we have not paid much attention to the restoration of the galley. This was not because it was not in need of a complete facelift (if only!) or because it was being overlooked. Some work has, in fact, already been done in this area but, leaving aside aspects of chronology, we thought it would be more illustrative to concentrate all the work done in the galley, step by step, into one single chapter.

Samba's galley, like those of most boats, old or modern, was based on prefabricated furnishings. When she was built, the shipyard put the kitchen together in their carpentry workshop and then installed it on board as a complete unit. With this kind of assembly it is far from easy to make later modifications to the layout or carry out partial repairs. Many of the screws or bolts used in the assembly are no longer accessible once the galley has been fitted into place. Essentially, you are faced with two possible solutions, one is to restore the existing base and the other is to start from scratch, ripping out the whole galley and redesigning it from top to bottom.

After weighing up the pros and cons of these two options in the end we decided to stick with the original galley and fix up the damaged elements as best we could, replacing anything that was missing or had finally given up the ghost after many years of service. As a whole, the work that we did on the galley is the part of this project that, technically speaking, most resembled restoration work.

The final result was satisfactory. Apart from the evident facelift, we now have a cold box, we have recovered quite a lot of stowage space and have maintained all of the worktop space, with which *Samba* was extremely well equipped for a boat of her footage, and we also installed a pressurised freshwater system, plus a seawater supply by means of a foot-pump.

Step by step

➤ *A reminder of what the kitchen looked like before we started. Although the base was pretty sound, the overall aspect was rather depressing. Some elements had disappeared over time, such as the doors, while others were on their last legs, such as the cooker, the hob, and the sink taps. There were also others in need of an urgent update, such as the cold box. Apart from the repair work, a good coat of paint and varnish is also a guaranteed solution when it comes to brightening the place up.*

➤ *As usual the first step was to strip out everything that could be dismantled until we arrived at a basis from which we could rebuild the galley. Without entering into aesthetic judgements, the rich variety of colours that previous owners had combined in the galley was, if nothing else, surprising, although obviously there is no accounting for taste. A case in point: no sooner had we clapped eyes on her than we decided that the synthetic pink-beige veneer of the furnishings just had to go. As this was not a crucial matter in terms of the refit as a whole, we kept on postponing it and the months went by. Then, suddenly, this 'seventies' colour once more came back into fashion, at least according to the decor magazines and, in the end, we even began to find it quite attractive, although who knows whether we will tomorrow.*

➤ *Doors and panels were taken off and taken away to be stripped and repainted. We cannot repeat too often how inconvenient it is to try to do this kind of work on board. You run a serious risk of leaving stains and the smells and dust that this kind of work always produces are difficult to get rid of. Beneath the coats of green, brown, and yellow paint we found a rich variety of other paint and primer colours, all of which had to be patiently removed, until we were down to the original synthetic veneer surfaces.*

➤ *Three good coats of satin-white soon gave the galley a rejuvenated look. On the edges of the pieces where the veneer was broken and/or had come unstuck we chose to completely remove the edging veneer, replacing it with three coats of varnish. In the photo you can also see how the main body of the furnishings had already been painted white, work that was done along with the rest of the interiors.*

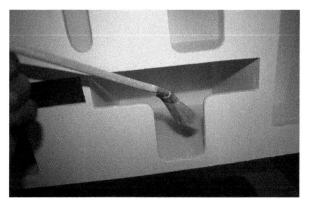

➤ When we did the main interior painting job, with Samba completely stripped out, we used a compressor and spray gun to ensure the best possible finish. But when you are dealing with the interiors of cupboards and lockers a spray gun is no use at all, only a brush will get you into all of those nooks and crannies, angles and side pieces. In the case of the shelves for plates and glasses, in the photo, things were even more complicated and we had to resort to a bent radiator brush, to get at the inside surfaces and back corners.

➤ The underside of the cold box lid, like the rest of the furnishings, had been covered with various coats of paint, and had become delaminated, with the insulation poking through all round the edges. After we had repaired the lid we gave it a complete coat of two-part polyurethane paint.

➤ The carpenter then fitted a new latch, to replace the primitive knotted rope that had previously performed this function.

➤ We also took advantage of having the loose lid to fit insulating seals. This was essentially a question of gluing on strips of rubber using contact adhesive. With the cooling unit already installed, the cold box was now ready to do its job.

➤ The part of the galley in worst condition was probably the worktop. The water had seeped in around the sinks and had rotted the plywood. At the same time we just had to get rid of the original plastic sinks along with their taps. The problem was that it was virtually impossible to actually remove the worktop as the screws holding it in place were inaccessible. The easiest solution, after removing the parts damaged by rot as best we could, was to fit the new worktop over the old one.

➤ A question of great importance in the case of Samba's galley was the positioning of the sinks. Constricted as the space was, between two structural bulkheads, there was no room for manoeuvre. In the end, however, we managed to find two sinks that fitted perfectly, installing the taps in the gap left by the smaller of the two. After taking measurements and adjusting the piece of plywood for the new worktop, the carpenter painstakingly marked out the position of the sinks and taps.

➤ On board a boat, whenever possible, it is always better to have two sinks rather than a single one, as whatever you put in either of them while sailing has less chance of falling when your boat heels over.

➤ We finally settled on a white synthetic veneer to line the new worktop, made of a material that is highly resistant to the knocks, scrapes and damp so typical of all galleys. We also painted the edges and the underside of the plywood base to protect it from water. We then fixed it permanently in place with three screws, inserted from below in the only accessible area (see the foreground of the photo) and a good bead of polyurethane adhesive for the rest, particularly round the edges, to avoid water seeping in.

➤ Once the worktop had been fitted into place all we had to do was eliminate the excess sealant using a piece of cardboard as an improvised scraper, finally mopping up the last traces with damp paper towels. If necessary, you can eliminate dry sealant using turps.

➤ We also glued the raised trim into place using polyurethane glue and nailing it into place using headless nails, which almost disappear from view under a coat of varnish.

➤ The next step was to install the sinks, which are also made firm with a little silicone around the edge of the holes. The new worktop, with its raised trim, and the new sinks and taps now look far better than they did before.

Two-piece hinges

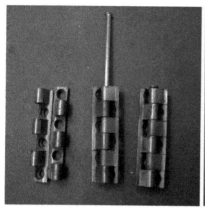

➤ For the galley cupboards, where the facing panel also acts as a stop for the doors, we used two-piece hinges, fitted together with a pin. These are quite common in normal household kitchens, and can be screwed into place discreetly along the edge between the panel and the door. You never find these hinges, which are on sale at any hardware supplier, in nautical catalogues. We used Zamac hinges, which is an aluminium-based alloy that will successfully resist a marine environment. The only inconvenience is that the chrome finish comes off quite quickly and oxidation darkens the metal. You also have to watch the oxide on the pin, which is usually galvanised iron.

Storage room door

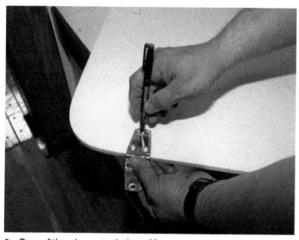

➤ One of the characteristics of Samba's interior layout is that, instead of the classic locker underneath the cockpit benches, there is a large interior storage room directly accessible from the galley. After stripping down and cleaning this door, instead of leaving it superimposed on the bulkhead, as it had been originally, we trimmed it down so that it would fit inside the frame, flush with the bulkhead, fitting it in place with exposed stainless steel hinges.

➤ A simple trick that we used to keep the a door in place while marking out and fitting the hinges was to wedge it in place with toothpicks, around the whole perimeter, so that the gap would be the same all the way round.

➤ We also replaced the latch, fitting a simple bolt. To cover over the holes left by the previous latch/ latches we used a piece of teak panelling as a base, left over from a neighbouring boat that was having its deck replaced.

Fold-away shelf

➤ One of the ingenious solutions used in Samba's galley, which we were delighted to maintain, was this fold-away shelf, which served as an optional extension to the galley's worktop. After stripping it down, cleaning it up and varnishing the edges, we spent some time trying to figure out how we could improve the system for locking it in place. The original shelf had an awkward crossways bolt that fixed it in place against the cold box when folded, plus a complex system of hooks that held it up in a horizontal position. In the end we found a system that was both simple and effective (the two qualities that best define any good solution). By attaching this single and solid bolt, the shelf can now be fixed in both positions with one single movement and mechanism, either folded away or extended, with the catch in this case attached to the engine housing bulkhead. A stainless steel plate reinforces and adorns the hole in the bulkhead.

➤ A simple trick that will allow you to paint or varnish pieces on both sides is to support them on nails embedded in pieces of wood, like a fakir's bed. This system is useful for items of any size, such as a cupboard door. First you paint the 'bad side', the side that will be less visible once the part is installed. Then you lay it down, supported on the tips of four nails, which will leave only small, barely perceptible, marks on the recently painted side. You are then free to paint both the 'good side', and also the edges, without having to wait for it to dry.

Renovating the burners

➤ Installing the new cooker was undoubtedly the most expensive and eye-catching part of our galley restoration, although it only took us a few minutes to do.

➤ Fitting the cooker in place with the corresponding gimbal mounts is as easy as making up the four screws that hold the mounts in place. In the case of Samba we previously had to make up a pair of wooden blocks to narrow the gap on either side.

➤ While the high price of nautical cookers bears no comparison to their domestic counterparts, there is no excuse for not changing old gas pipes and valves in which you no longer have any confidence. For a small investment in both time and money, less than an hour's work, you can replace all the gas piping and valves, plus the gas bottle and regulator valve on any boat. Gas is a serious business.

➤ Comparing the present, rejuvenated, aspect of the galley with the first photo in this chapter shows the improvement that we have achieved, both from an aesthetic and a functional point of view.

Updating the electrical and electronic installations was one of the most complex aspects of Samba's refit. In the end, like an iceberg, hardly any of this immense task of engineering would actually be visible, hidden away behind the furnishings.

Finishes and testing

The electricians first started to do their work when we had the interiors completely stripped out. They drew up the wiring diagrams, decided on the location of each element, and started to install the wiring and connect up some of the accessories (batteries, converter, charger, etc.). Work gradually continued on the electrics and electronics as the refit in general advanced, with changing the engine, the installation of the freshwater and bilge pumping systems and as the furnishings were gradually fitted into place. Finally, after a series of stages, staggered over a period of time, now that the ceilings and walls had been lined and with Samba's refit almost finished, everything was ready for the electricians to conclude their part of the work.

Because of the extended time scale involved, a lot of planning had to be done. In questions of electricity or electronics you cannot trust to your memory alone; you need to base the work on a wiring diagram so that you can pick it up again at any point during the refit without any questions.

Connecting up a ceiling light fitting, a bilge pump or even the plotter/GPS at the chart table, all part of the final goal and the most visible aspect of an electrical refit, is also work that is, generally speaking, within the grasp of most amateurs. As we have mentioned on various occasions, the engineering of the electrical and electronic installations are much more complex than simply fitting everything in its place.

One of the least pleasing aspects of renewing these installations is that the most laborious and sophisticated part is precisely the part that never gets to be seen. While the painting or carpentry work stands out in all its splendour once completed, the hundreds of yards of main and branch wiring installed on board are only evident when it emerges at the light fittings or control panel. Everything else, the most important part, is all hidden away, out of sight behind the furnishings.

Anticorrosion: Titanium anodes

In order to avoid the corrosion of underwater metal parts, we chose a system using titanium anodes. Galvanic corrosion is caused by the flow of electrons (mass) between different metals due to the minute electrical current induced by any metal in contact with water or connected, even if indirectly, to the boat's electrical field. This problem is aggravated by the progressive sophistication of the electrical and electronic installations on today's leisure craft. Looking for the precise origin of this corrosion is very complicated, insofar as it may be found in any combination or interrelation between the electrical equipment, other metal parts of the boat and the affected metal. It is also possible

that this problem is the result of defective electrical installations.

The usual solution is to use zinc anodes. This metal is more open than steel or bronze to releasing its electrons, which means that it is the anode that suffers corrosion, preventing the other parts from becoming degraded. By installing and regularly checking anodes in the rudder, the propeller shaft and the keel, all these parts can be protected. Anodes will not, however, prevent galvanic corrosion from affecting your tanks, exhaust manifold or through-hulls.

This question is made worse by the fact that when boats are moored up at a jetty, and permanently connected to the 220 V, 240 V, or 110 V grid (depending on whether you are in Europe, UK or USA). In some way they form a kind of galvanic cell, via the onshore earth. A boat with a metal hull and/or one that is scarcely protected, can suffer a much higher than normal rate of corrosion in its neighbours.

The heart of a titanium anode anticorrosion system is a regulator to which all of the metal parts to be protected must be connected. The regulator controls the overall electric potential of these parts (through-hulls, tanks, shaft, rudder, keel, etc.) and automatically counteracts it in a controlled way in the water via titanium-metal anodes, that are much less toxic than zinc, and are installed as simple through-hulls. The system has a monitor with LEDs, arranged like a traffic light, which indicate the level of protection in the boat at all times.

Step by step

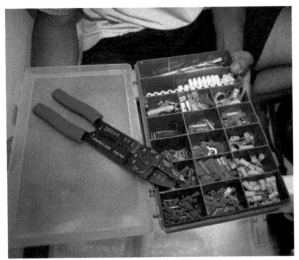

➤ *Clips and connectors, on sale at any hardware store, are both essential for the finishing of any electrical installation. Other essential equipment for an electrician, whether professional or amateur, is insulating tape, self-vulcanising or thermoretractile tape, for connections that are more exposed to humidity, and a small screwdriver. Assembling the necessary kit will hardly cost you anything at all.*

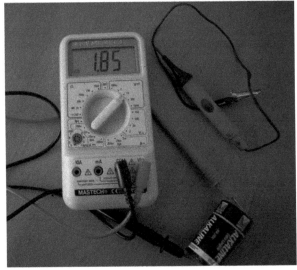

➤ *It is also important to get hold of a voltage tester so that you can analyse electrical flows. The first question to be asked of every connection is always: "Is it live?" If you do not have a voltage tester then you should at least rig up a system with a bulb and a couple of wires that will enable you to find out whether your connection is live or not.*

➤ The electricians came and went as our refit advanced, completing the installation and testing the equipment as it was connected up (engine, pressurised freshwater, cold box, anchor windlass, etc.).

➤ Thuroughness and order are fundamental from start to finish in any electrical installation. You would not have to have been an expert to realise that Samba's old installation, with its mishmash of wires, connected up every which way to the back of the switchboard could not, by any stretch of the imagination, be reliable. In contrast, the neat and well-ordered strips of the new panel give a completely different impression.

➤ During our refit we also replaced the old, broken down, compass. To protect the electrical wire that runs up the pedestal, and to avoid it getting snagged in the rudder mechanism, we ran it through a length of corrugated hose attached to one side of the pedestal.

➤ Splicing wires together is something that must be done seriously, using the right connectors and quality strips or fasteners, as the case may be. It is a shame to spoil an electrical installation with connections of doubtful reliability, even more so when they involve hardly any cost at all. In some cases, a little insulating tape is enough. Where there are damp conditions (in bilges, etc.) you should use self-vulcanising and/or thermoretractile tape, while in very exposed locations, or where access is more difficult (mast or exterior connections), you should also solder the tips of the wires. When in doubt it is better to be too cautious rather than too confident.

Titanium anodes

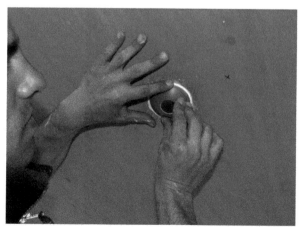

➤ Two titanium anodes, which look just the same as through-hulls and are installed in exactly the same way, ensure the protection of your boat against galvanic corrosion.

➤ All the metal parts of the boat, from its keel to the through-hulls, including the rudder and any metal water or fuel tanks, are all connected up to the regulator, which controls their electric potential.

➤ To protect the shaft, spring-loaded clips are fitted. These will ensure that there is permanent contact and will protect against corrosion, whether your boat is at sea or in port.

➤ The regulator and the control panel are the brains of this system. The needle indicates the equipment's consumption, in milliamperes. When you press a button an LED light appears, indicating the level of protection: if the light is green, the boat is correctly protected; if it is red or amber, the boat is respectively either under or over protected against induced currents.

220 volt installation

The refit of Samba's electrical installation included a small independent 220 volt circuit, with four sockets (chart table, galley, saloon and head), a port plug connected to the battery charger and a 12:220 volt converter. A 220 volt circuit on board a boat requires a number of precautions to be taken, starting with a tripping device that will disconnect it immediately in case of any problem, plus sockets located as far away and above the bilges as possible. The 220 volt circuit should be left unconnected when it is not being used, particularly when you are at sea.

➤ We installed the 220 volt circuit sockets (in the photo an exterior model with cover) as high as possible and with connections protected against the damp. In any case avoid touching them when wet or barefoot.

➤ Connecting a light fitting to the ceiling, a bilge pump or the GPS are all tasks that are very much within the reach of most amateurs. However, unlike other nautical tradesmen, the best electricians have to undergo top-level technical training. The sophisticated blend of electronics, information technology and electricity these days means that, even on smaller boats, an advanced level of knowledge is essential.

Professional finishing

One of the differences between electrical and electronic installation work done by professionals compared to amateurs is in the quality of the finishing. While the neat ordering and clamping of wires, along their whole length, in an orderly and discreet manner, with the use of the correct terminal connections will rarely affect whether or not an apparatus works correctly, at the time of its installation, this is not a simply cosmetic aspect, as many amateurs tend to believe. Usually it is only in the medium and long term that a defective installation starts to become apparent. So many of the electrical problems (malfunction, voltage drops, wires getting nipped, etc.) that professionals are always getting called out to fix are simply due to slapdash installation work.

There are few things guaranteed to spoil the look of a boat's interior more than having electrical and electronic wiring running all over the place (in corners, across bulkheads, over ceilings and/or shoddily covered over with insulating tape). Shipyards are fully aware that, despite the fact that owners demand ever more complex electrical installations, the last thing they want is to see a single wire. To achieve this aim, above the false ceilings, these days you can purchase prefabricated bundles of wires, with the individual connections only emerging at either end to be connected up to the different pieces of equipment.

While this system provides exceptional neatness, it can also turn an occasional repair into a nightmare. In the case of *Samba*, apart from the fact that she was built before false ceilings had been invented, we preferred to install the wiring hidden away behind the cupboard and lockers yet, at the same time, always accessible. This may not be the most elegant solution but, as most electricians will testify, it is one of the most convenient in terms of accessibility for inspection and dealing with any problems that might arise.

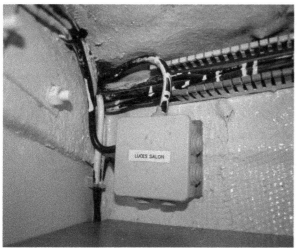

➤ *To avoid problems of accidentally crushing or pulling out the wires of your installation you have to ensure that they are neatly aligned, firmly clamped and/or enclosed in wiring ducts. You should always use connection strips, rather than exposed connectors, and the installation of a junction box is also recommended.*

➤ *In some cases there is no easy way to hide the wiring. For example, for the interior connections of the mast wiring we had to install a discreet duct, adapted to the colour and contours of the ceiling.*

➤ *For this light fitting in the forward cabin, there were 10 cm of exposed wiring that could not possibly be hidden. Instead of black wire, we used a white protective sheath, blending it in with the colour of the bulkhead.*

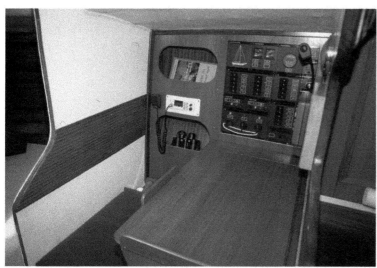

➤ *The chart table is the neural centre of the great change that we achieved through the renovation of Samba's electrical and electronic systems. Not only have we greatly improved the look of it, we also improved a number of technical aspects (charger, converter, autopilot, electronics, etc.) and have made an immense leap forward in terms of both reliability and safety.*

The worst part of the work that we put into the electrics and electronics is that, like the tip of an iceberg, the greater part is inevitably hidden from view. By the time we had finished, after the hundreds of hours' work put in by the electricians, the only visible evidence of all that effort were the light fittings, the electronic repeaters and our new instrument panel. Everything else is neatly hidden away behind the furnishings.

Chapter 41

Sails

When buying a used boat, at the bottom of her lockers new owners often find piles of old sails, of all shapes and sizes, origins and conditions.

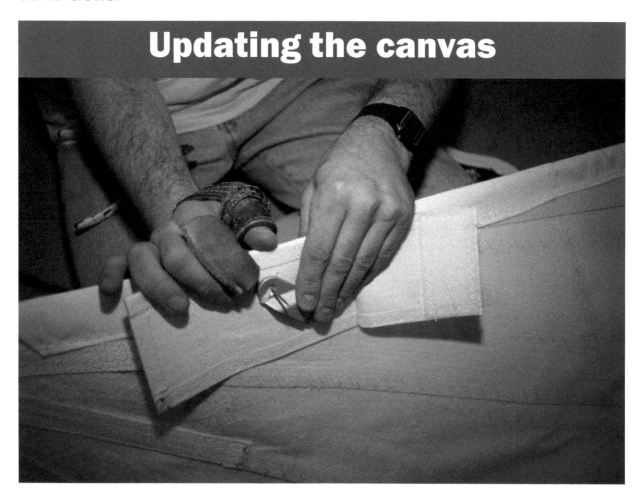

Updating the canvas

That was exactly what happened to us anyway, and the first thing we did was to take it all to a yard, lay it out and make up a complete inventory, deciding on what could still be used and what was in need of immediate retirement.

During this first selection it soon became obvious that none of the genoas were going to make it through. But the fully battened mainsail and the spinnaker (an old but little used tri-radial) were still in reasonably good shape and quite well preserved. Apart from these two, only the spitfire, the storm mainsail and perhaps one of the storm jibs were even deserving of the description

'sail'. The rest were only of use as Dacron awnings.

When we bought *Samba* we already knew that she had a canvas problem and had accepted the idea that we would have to replace her headsails. The range of combinations facing sailing enthusiasts when it comes to choosing new sails, cloths and cuts, is immense and can only be dealt with by a kind of elimination based on footage, sailing programme, where you expect to be sailing and the owner's own particular taste. All of these factors will have their influence on the choice and, between them ought, to ensure the best set of sails for each case.

So, given that our aim was basically to do a bit of cruising, a furling genoa was the most logical option for our main headsail. We had already rejected the possibility of more complex laminated sails, because, at the time of the renovation, these were too delicate for furling and their expense could only really be justified for racing. However, technology has now moved on and nowadays there are furling laminates called 'cruising laminate sails' which are strong enough and widely used for cruising. Returning to our renovation, given the large surface area of the genoa (representing 60 of the boat's 85 square metre sail area), plus the North Wind's ability to sail close-hauled, in the end we decided on a Pentex genoa (made from Dacron) with a 150% LP and with radial cut. This would be the boat's principle driving force on 85% of the days it would spend at sea. It was well worth concentrating the investment right there on that sail!

However, although there are no problems with a furling genoa in gentle breezes or steady winds, when it begins to blow a bit, it can become difficult to maintain a decent sail shape, above all when sailing close-hauled, which is when your boat will start to heel over and fall to leeward, making it very difficult to fetch her to windward. To cover this eventuality we ordered a Dacron Solent (100% LP) with horizontal panels and reefing bands that allow it to be reduced to the dimensions of a forestay.

Succumbing to modernity

Sailmakers continue to be a sanctuary of craftsmanship, although these days they also depend to a great extent on computers and modern manufacturing processes. The design parameters are no longer locked inside the heads of the master sailmakers but are available in computer programmes that will allow precise cutting instructions to be repeated over and over again for each panel of each sail.

The cloths used (in terms of cruising or cruising/racing sails) all tend to be much of a muchness. In the end, the difference between sails is not so much their design, cut or cloth, rather it is in taking advantage of the enormous ability of computers to adjust each sail to a given boat and that boat's specific sailing programme.

The price per metre of cloth can range, depending on its weight equivalent, over twenty times between polyester and the most sophisticated laminates. But skilled labour is always going to be a significant part of the cost of any new sail, and this cost will increase in direct proportion to the sophistication of its making up (radial cut, reefing bands, full battens, glued seams, etc.). Choosing the sails you want for your boat and the sail loft that you want to make them up is, therefore, a question of finding the best compromise between all these concepts, starting with the reputation of the sailmaker and followed by the design, quality, specifications, durability, price and (let's not forget)

after-sales service that each company is prepared to provide. It is not always easy to get this equation right.

Cruising sails: cloths, laminates, cuts, stitching and gluing

A sail is essentially an aerodynamic profile that has to remain unalterable before the wind. The advantage of the most expensive and sophisticated laminates is that distortion is minimal, allowing for the use of finer 'cloths' for the same wind force. To put it another way, you could get a Dacron sail made up with the same resistance as one of your top line laminates, but it would be spectacularly heavy, an unforgivable aspect for racing and also one that would impose serious restrictions on a cruising boat such as *Samba*.

Firstly, then, we divided up the choice of sail on the basis of the materials used, between standard cloths and laminates. After that you have to decide on the cut and the making up. Your sailmaker will be able to guide you on which material to use.

Cloths: A cloth consists of yarn woven together to form a compact surface. For sailmaking the most common cloth is Dacron, although Pentex (a derivative of Dacron), Spectre and Kevlar are also used. The weak point of a cloth is what is called the 'bias', which is when it is stretched diagonally. A sail, by definition, produces forces in triangulation, from its three corners. This means that when your sails are being made up, the weight of the cloth has to be over-dimensioned to ensure that the sail will not be pulled out of shape when subject to these angled forces.

Laminates: A laminate consists of two layers (or laminates) glued together and imprisoning a screen of threads between them. The external layers are generally Mylar or taffeta, which is a fine layer of Dacron. Mylar is lighter and has more stretch resistance, but is also more expensive and less resistant to use than taffeta. Just about every other day new combinations are developed for sandwiching between the two outer layers, with varying thicknesses and alignments of different yarns (Kevlar, Spectra, Pentex, Twaron, carbon, Vectran, PBO, etc.). These yarns, generally referred to as high-performance yarns, cannot be woven, in the strict sense of the word because, in tightly packed sections, they tend to break or lose their properties.

Cutting and making up: As the interior yarn is aligned with the anticipated forces acting on the sail, laminates can handle heavy winds without losing their shape at low sail grammages (thicknesses). At the same time, modern 'making up' systems, with sophisticated moulds and / or the gluing of panels, persist in this slimming cure. But if any investment is worthwhile on a racing boat in order to shed a few pounds, or even a few ounces, a cruising boat has to carefully weigh up and evaluate such extra costs. It must also be said that weight differences do not increase with the length of your boat on a mathmatical basis, but on a geometrical one. For a 30-footer, the weight gain of a Dacron mainsail in comparison with a high performance one is barely 10%, but for a 60-footer it could be as much as 50% of sail weight, a completely different kettle of fish altogether.

In any case, and in the same way that cotton eventually disappeared following the invention of Dacron, everything seems to indicate that this material will, sooner or later, cede ground to the high performance laminates, which have already become essential for racing yachts and also progressively for cruising yachts.

Sails: usage chart

The scale of wind speeds anticipated for the use of *Samba*'s new headsails is summarised below:

0 - 5 knots: With indifference to the protests of purists, *Samba* will be using her so-called 'diesel genoa' to maintain a cruising speed of around 6/7 knots in windless conditions or when only the slightest breezes are stirring.

5 - 20 knots: As soon as wind speed gets up to over 5/7 knots the North Wind 40 will start to respond to the genoa at 150%, and can hang on under full canvas up to apparent speeds of about 15/18 knots. A speed of between 8 and 15 knots can more or less be defined as usual wind conditions in the Mediterranean (where *Samba* will sail once refitted) so it is well worth investing in sails that will allow us to take full advantage of it. By the time we get to 20 knots apparent wind we can start playing with the first turns of the furler and/or a reef on the mainsail.

20 - 30 knots: At these apparent wind speeds, although a furling genoa could just about hang on in there, the solent is our best option for sailing close to the wind, relieving the genoa of its stresses and allowing the option of taking in another reef when we get her up to 27/28 knots.

30 knots and over: In these wind speeds the best option available to your average cruising yacht is to forget about sailing close to the wind and look for shelter in the nearest port. If this is not possible, the storm jib or even the spitfire should still allow you to sail safely.

Step by step

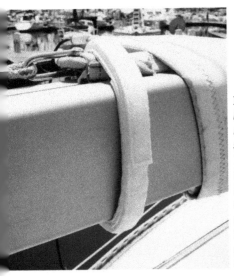

➤ The mainsail, with full battens and quite recently made, was still in good shape and has been kept. To avoid the slide damaging the track on the new boom in the medium term, as was the case with its predecessor, we installed this simple system with a grip and Velcro at the sheet corner.

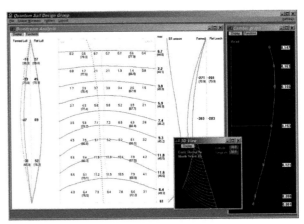

➤ The data collected was then entered into a computer programme. Working from the three corners of the genoa, for example, the design program established the shape of the sail, displaying its permanent parameters on the screen. These adjustments, to a greater extent than either the cloth or the cut, would be the cornerstone of the behaviour and 'character' of the sail itself.

➤ Before starting with the cut and making up of the new headsails, the sailmaker ran a thorough check of all measurements, hull, mast and rigging, plus references with regard to the type of hardware she was equipped with and how it was laid out. All of this data was entered into a file, which means that we now had a full inventory of the boat for any future sail requirements.

➤ The cutting plotters, used both by sail lofts and clothing manufacturers, work on large tables on which cloth has previously been extended. Following the instructions given by the computer, these machines cut the panels to precise shapes and alignments.

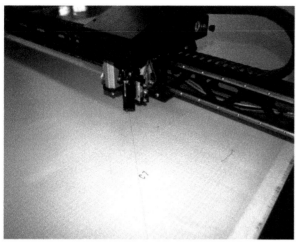

➤ The program itself made sure that the different panels were lined up to minimise cloth wastage. The information allows for millimetrically accurate repetition and the optimising of the shape of a sail as often as necessary, without the need to depend on the human factor.

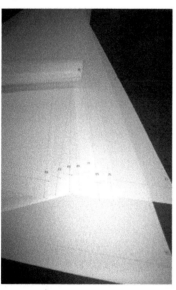

➤ The panels were individually numbered as they were taken from the cutting table, to simplify their future assembly, with clear indications and margins for gluing and sewing.

➤ The next stage, which was both slow and laborious, was to join the different panels. We used a double sided adhesive tape, faithfully following the line drawn out by the plotter.

➤ Once the panels were joined together, the sail was passed on to the sewing stage. Sitting down, for the sake of working convenience, at floor level, the operators sewed over the adhesive tape, joining the panels together following the indications of the plotter.

➤ In order to ensure the correct adjustment of the sail, in radial cut sails, each corner was sewn separately. Before they were finally sewn together, a check was run to make sure that the line drawn by the plotter was uniform on each of the panels and that they had not been misaligned during their gluing or sewing stages.

➤ This adjustment verified that the shape of the sail was as anticipated in the design program. The sailmaker used a felt tip to go over the continuity of the line between one panel and the next. To this end, he used a long rod, bending it until it met the design lines. The sail was then sent back to the sewing machines, where the three corner patches were added.

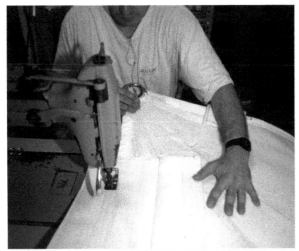

➤ Once the corner patches were made up, the sail, now in its final shape, was passed on for finishing. This is one of the few processes in sailmaking in which classic sailmaker's tools are still used, such as eyelet punches, awls and sewing gloves. Here they fitted the eyelets (cloth reinforcements at the corners already attached in previous operations), the leeches and telltales.

➤ Before bagging the furling genoa, the lead sailmaker laid it out on the floor for a first and meticulous visual examination.

➤ Following the first examination at the sail loft the same was done on board Samba, when the sailmaker came aboard to bend the genoa and the solent for the first time, making sure that everything fitted as it was supposed to, from the profile guide to the sheet corner.

➤ Taking advantage of an almost windless day, we raised the genoa to get a good look at how it hung before furling it away in the furler. Samba was now ready to go to sea with all her canvas aboard.

Chapter 42

Interior carpentry (part 6)

The business of adding finishing touches to the carpentry work seemed as if it would go on forever. While the excitement of having the end of the refit in sight was a stimulus, it was also a cause of added stress.

Never-ending finishes

One of the most commonly recurring issues in refits done by amateurs is relaxation. People will start a project with loads of enthusiasm, constant dedication and a reasonable budget, but then, little by little, the level starts to drop, they start taking technical shortcuts and some jobs keep on getting postponed or started and never finished. The result of all this is a boat that is permanently 'under construction', half done or half undone, depending on how you look at it.

It is very easy to lose your motivation as the finishing line gets closer, as it also is to throw in the towel before you have even got halfway and when the problems (technical, financial, etc.) start to multiply and you cannot see how you are ever going to get over them.

Knowing the enormous tenacity that is required to bring a large scale refit to a successful conclusion, all I can do is encourage any of you who have the bottle to attempt it and recommend that you do not drop your guard until you finally have the boat ready. Once you have packed away your toolbox it is really hard to get it out again the following winter to finalise all those unfinished jobs. You have to try to persevere with the finishing work with the same enthusiasm as in the early stages of the refit.

Step by step

A new cover for the engine compartment

➤ When it came to the engine compartment cover we had to start from scratch as we had thrown the old one away. True enough, the one we threw away was broken, difficult to use and no longer fitted over the new engine, but it is always a good idea to have something to base the design of any new part on.

➤ Luckily, the measurements were more or less rectangular and the carpenter was able to redesign it without too many difficulties. The new engine, despite being almost 20% more powerful than the original, was virtually 20% smaller in terms of its external dimensions. This allowed us to gain 15 cm of floor space and also to lower the height of the cover fitting by 7 cm.

➤ After we had made up the basic cover we were faced with a dilemma regarding the steps. On the one hand we would have liked to keep the old teak steps with their unmistakeably retro air. The photo shows them in place in what would be their natural position. Here you can see the old attachments on the floor, giving an idea of the space that has been saved as a result of the reduced size of the new engine.

➤ The problem was that the old steps were very narrow and difficult to use. As a result we finally ended up by attaching a couple of 'Japanese style' steps to the front of the cover.

➤ We adopted a rather unusual solution for fitting the cover in place with these lever fittings, originally designed as hatch fastenings. The asymmetric adjustment, which can be regulated with silent block, is similar in concept to a bicycle's handlebar or saddle height adjustment levers, and is far more effective than the usual braced bolts, which always end up leaving a gap.

➤ When all of the adjustments had been finalised the carpenter took the frame away to fit the moulding adornments, cutting them at an angle and gluing them into place with polyurethane adhesive. For the clamps he only applied sufficient pressure to hold the mouldings firmly in place.

➤ We then gave the whole assembly three coats of two-part satin varnish and, as a non-slip surface used plastic flooring tiles, which are very hardwearing and effective. First we cut the pieces to size and then glued them in place using polyurethane adhesive. The mouldings, when attached, also served to cover over the heads of the screws that we had used to fit the cover together, enhancing its appearance.

Restoring the companionway

➤ The companionway doors were in a bad way, with the varnish cracked and dulled by the sun's ultraviolet rays, the joints had dried out and the lock was broken. As always the first thing we did was strip and clean until we got it down to a solid base. After removing the lock we then scraped out the black joints between the decorative teak planking on the exterior surface of the doors. To do this we used a plastic scraper (sold at any hardware store), although a screwdriver or small chisel will do the same thing. The purpose was to remove all traces of the old, dried-out joints.

➤ Once we had done this we went over the joints with a polyurethane filler, the same as used for deck joints, which is resistant to both the sun's ultraviolet rays and seawater. Then we applied a coat of transparent primer, to make sure the sealant would adhere to the old wood.

➤ The sealant was then liberally applied. As you can see in the photo, we made no attempt to limit this to the joints.

➤ Almost two weeks' later, allowing a good margin in terms of the five days recommended by the manufacturer, we could now take it for granted that the sealant had set. This can initially be assumed when the surface is no longer sticky to the touch, nevertheless we would recommend that you wait much longer so that you can be sure that it has set all the way through and will not break off when sanded down. If this happens you simply have to go back and start all over again. The advantage of starting at the joints is that, as you sand off the excess sealant with an orbital sander, you will also take off the varnish. In less than an hour we had the teak looking as good as new.

➤ While we were working on the doors was the perfect time to sort out the door frames, where the varnish had also dried out and blackened, following almost thirty years of sun and salt. Several coats of stripper, sanding and scraping were necessary to get down to the wood. When using paint stripper aboard it is important to make every effort to protect the areas around where you are working. You also have to leave a margin of time (at least three days) before applying the new varnish, paint or oil to the wood. If you do not, the effects of the stripper remain latent in the pores of the wood and could easily spoil any new coats you might apply.

➤ We also took advantage of the occasion to repair a part of the frame that had split. When wood starts to split it is usually better to break the split piece off completely and stick it back on using a coat of glue on both surfaces. You can then use clamps to keep the part firmly in place until the glue sets.

➤ *The next problem was levelling the interior companion frame with the ceiling panels. When laminating a plywood reinforcement to the underside of the cabin roof we had to sacrifice a few centimetres of interior headroom. As a result the frame was no longer flush with the ceiling panels. As the old frame was in a pretty bad way the easiest solution was to get rid of it and make up a new one from scratch.*

➤ *In order to save ourselves the job of machining 3 metres of solid 10 x 12 cm teak battens and then having to individually cut and adjust each of the parts to be installed, we decided to fit a false frame. As a base we used polyurethane adhesive to fit lengths of plywood round the frame, leaving a 1.5 cm. space all around where we would subsequently attach the strips of teak as a visible finish. The nails you can see hammered into the plywood are an easy way to position these lengths with force and precision without getting polyurethane glue all over you.*

➤ *With the false frame in place, our professional carpenter took over to fit the teak finishing. This is another fine example of the advantage of combining amateur and professional work. Every enthusiast ought to know his technical limits and when the time has come to hand over to someone with the necessary skills. Doing it this way also helps to lower costs without compromising the final result.*

➤ *In one morning the carpenter managed to precisely fit the teak strips in place, gluing them in an L-shape with polyurethane all round the frame.*

➤ *Once the glue has set, any excess can easily be eliminated, first with a scraper and later by sanding down, gradually using finer grade sandpaper.*

➤ *The time had also come for the companionway's old handholds to take their well-earned retirement. Rather than trying to restore them, particularly in view of how little they cost, we decided to replace them. For items that are going to be subject to some force, it is important to drill the holes for the attachment screws precisely and, above all, perpendicularly. To ensure that this would be the case, we clamped them firmly in position and then drilled the holes through the frame.*

➤ This is what Samba's companionway doors now look like, after the wood received three coats of teak oil. On the outside we disguised the hole left by the original lock by fitting a plastic air vent in its place. We also decided on a latch and padlock system for locking the doors, possibly not as refined as an embedded mortise lock but far more practical and long lasting. We had the stops for the sliding companion top made to measure in stainless steel, after first making up cardboard templates. This system is much more discreet than the plate and chock studwork that had previously been employed.

➤ The look of the companionway has also improved immensely from the inside. The new frame fits perfectly flush with the ceiling, while the white painted finish of the sliding hatch is a vast improvement on the mishmash of colours that it had previously worn.

Soundproofing the engine compartment

Soundproofing the engine compartment is essential in terms of liveability on board. It is also, technically speaking, quite easy to do and, in the case of Samba, we had the opportunity to experience, step by step, the improvement that it made. From the first time we started up the engine to the completion of the soundproofing of the cover there was a spectacular reduction in decibels. What had previously sounded like a power station working overtime was reduced to a distant murmur, barely perceptible from outside the boat. Originally Samba's soundproofing had consisted of the classic honeycomb-foam, which we got rid of straight away. Due to its low cost this foam is still used today by a number of shipyards, although it is actually quite limited in terms of acoustic insulation and has the added disadvantage of absorbing diesel fumes, which means that, in the case of fire, it tends to go up like a torch.

➤ For both safety and comfort we preferred to invest a little more and install a triple layer of 25 mm thick foam, sandwiched between layers of lead to reinforce soundproofing qualities.

➤ As we did not completely trust the self-adhesive film as a medium-or long-term solution we reinforced it with strategically placed screw-on plastic fittings to avoid the panels peeling off.

➤ These panels, as well as notably reducing the decibels, act as a flame retardant. Installing these panels was easy enough, using a jigsaw, a cutter or scissors we managed to trim them to the required shape and then fit them into place around the engine compartment using self-adhesive film. Little by little we lined the whole of the engine compartment.

➤ For the soundproofing of the cover we started off by marking out the line of projection of the soundproofing panels alongside the beds, or where the inspection hatches and bulkheads fit together. Later, with the cover on dry land, for convenience, all we had to do was cut and fit the sections in place.

➤ After checking again to make sure that everything fitted correctly into place and that none of the engine parts, particularly moving parts, came into contact with the panels, we gave the compartment a final professional finishing touch with duck tape round the exposed edges and corners of the foam panels.

➤ Originally, in the storage locker, there had been a complex fold-down inspection hatch providing access to the engine and the stern tube. In its place we fitted a simple cover, over a frame of strips made secure with a couple of sliding bolts. As always in these cases, the shape of the hatch cover was only nominally rectangular, and first of all we had to make up a template to get the shape just right. A couple of coats of paint and a plastic grill, to allow the hot air produced by the engine to escape, completed the visible side. On the back we glued a single width panel of soundproofing foam into place, the same as we used to line the roof of the engine compartment.

➤ The soundproofing of the engine compartment was then completed with another hatch, located in the aft cabin. For this trap we used the original part, which had been cleaned up and lined along with the rest of the interiors. Previously the latch fixing this hatch in place had been a rough and ready wooden peg, which we decided to replace with a flush fitting catch, identical to the ones that most production boats are now fitted with. Inside, and in order to reduce the sound in the aft cabin as much as possible, we again used a triple thickness foam panel.

As amateurs and without any previous experience in this kind of work, we managed to soundproof the engine compartment, its cover and the inspection hatches in barely 5 hours.

It is the finishing that defines the quality of the work you do. And this is true not only for carpentry; the same can also be said of painting and varnishing, and the final details of your electrical or water installations.

Attention to detail

Before starting any job on board, no matter how trivial, it is important to make a list of everything you are going to need, because if you do not bring it all on board with you, you will end up losing absurd amounts of time running backwards and forwards to the hardware store or marine chandlers.

The work illustrated in the present chapter was done by a number of different people, including the owner and expert professional carpenters. While the level of difficulty of any job is a relative concept, everybody has their technical limitations and it is a mistake to blindly rush into a job that is beyond your abilities. Doing so is a virtual guarantee of a bodge.

To change the subject, one of the advantages of wood is that a variety of tones (pine, sapele, mahogany, cherry, teak, etc.) does not necessarily look bad and on occasions may even look more attractive. But before launching yourself willy-nilly into whatever comes to hand it is a good idea to check out your combinations and see how well they are going to work. This 'presentation' should be done with pieces of wood that have been either varnished, or soaked in turps (the optical effect is identical) as the colour of the wood will change with the application of a finishing coat of varnish.

All woods, including teak, have different tones depending on their country of origin, the size or variety of the tree, plus a whole series of other factors. Shipyards, to avoid colour mismatching, try to make sure that a single batch is used for the carpentry on each of their boats. Of course with a partial refit, where you have to combine old and new woods, this

is impossible.

For *Samba* we used teak as the basis for the carpentry, although other types of wood are also acceptable for use on a boat. Teak can be difficult to find in DIY stores although it is not exclusive to marine carpenters either. During the refit we had different non-marine professionals supply and/or work this wood for us. It is worth doing a price comparison.

Step by step

Two new doors

Delivered in kit form some thirty years ago, *Samba* must originally have had three interior doors. One for the storage locker next to the kitchen; another, which we had decided to do without for the moment, to the aft cabin; and a third, the restoration of which is shown in the following photos, between the head and the forward cabin. Strangely enough, on *Samba* there is hardly any separation between the saloon and the head (before there was just a curtain), another aspect that we will be putting right in this chapter. We have simplified the restoration of the doors by dispensing with overlapping frames and installing them flush with the bulkheads.

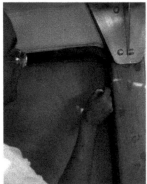 

➤ *To draw a door inside its frame, the simplest thing to do is to trace its outline on a panel held up against the doorway. Ideally you should have somebody else to hold it in place from one side while you mark the outline from the other. Two 5-mm thick strips of wood are perfect to allow an adequate gap at the bottom to ensure the door does not scrape the sole when opened or closed.*

grain. Another good idea is to use saw blades for cutting metal. Although the fine teeth of these blades will mean that the saw advances more slowly, the cut will be much cleaner than it would be using saw blades specifically for wood. It would be rather unusual if your door fitted immediately after the first cut; you will generally find that you have to sand down the different edges until its fits neatly into place on all sides. In cases where the door overlaps an L-shaped frame, a standard fitting in production boats, the adjustment will be easier to manage as the frame will cover any gaps. In our case, as the doors had to fit flush with the bulkheads leaving any gaps perfectly visible, the adjustment had to be precise.*

➤ *Once the finer adjustments to the door had been made, it is time to fit the hinges. In the case of the forward door of the saloon, which swivels on the mast pillar, we used flush hinges that we slotted in between the door and a strip screwed onto the pillar, which helped to widen the opening angle of the door.*

 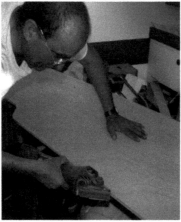

➤ *Once the outline has been marked on the sheet of plywood all you have to do is cut it out, following the line with a jigsaw. To stop the sheet from splintering we used paper tape, particularly when we were cutting across the*

➤ *For the access door to the forward cabin we used these flap hinges, screwed into place. These are the simplest kind of hinges for doors that close flush with the bulkhead.*

➤ For both doors we used turn-latch type fittings, with their exterior mechanisms bolted to the door. For these kind of latches, all you have to do is drill a hole in the door to pass the shank through, avoiding the need for internal slots which can be complicated to install in doors that are not very thick (ours were 13 mm). We then fitted a catch directly to the edge of the frame (supplied with the latch). The last stage, obligatory in both cases, was to fit stops, to avoid the doors swinging right through the plane of the bulkhead.

Finishes in the forward cabin

➤ The finishing of the edges and separation of the bulkheads in the cupboards of the forward cabin, using U-shaped (triple-angle) wooden strips, was sufficiently complex to leave to the professionals. By cutting, sanding and adjusting the strips with accuracy and precision the carpenter managed to get this done in little more than one morning's work. We cannot stress enough the advantages of alternating your own do-it-yourself skills with those of professionals as a way of controlling your budget without sacrificing quality. In the finishing stages, for example, we only had the carpenter on board for three days, but none of his time was wasted on tasks that we could handle ourselves. Often, if you overstretch yourself in a technical sense, this will end up in more hours for the professional, who then has to spend time undoing whatever you have done badly before he can do it as it should have been done in the first place.

➤ Once we had decorated all the exposed edges in the cupboards and along the bulkheads, the next step was to glue sections of 5 mm teak-veneer plywood sheeting onto the sides of the cupboards, to cover over the imperfections left in the bulkhead by the original berths that used to occupy this area, traces of which can still be seen inside the cupboard.

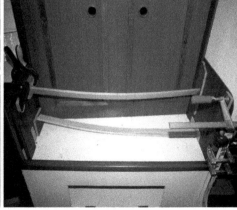

➤ First of all using a saw and then with a sharp knife, the carpenter meticulously cut these sections to size and adapted them to this decorative function. After varnishing the mouldings we glued these panels into place using polyurethane glue, with clamps and bracing strips of wood to hold them firmly in place until the glue had dried.

Skirting boards: taking care of the details

Once all the furnishings had been installed and the interior linings fitted, *Samba* still needed a series of mouldings to neatly finish off a number of edges and the borderlines between different materials. The work involved in doing this is so laborious that production boats now tend to avoid it altogether. These days furnishings are designed in modules that do not require any subsequent finishing.

The mouldings were installed by either gluing them into place or using visible screws. We dismissed the option of blind screws with a wood finish cap, which is much more elegant, on the first day. Apart from the fact that our budget did not stretch to such subtleties, blind screws almost always require the varnishing to be done on board, which is a nuisance.

Once we had decided on our general approach and put in our order for the required lengths of moulding and skirting board, installing it was easy enough for us to do ourselves. A fine-toothed saw, a sharp chisel and a hand plane were the only tools we needed. For purposes of fine adjustment, a file and/or even a cutter may come in useful, plus sandpaper, is of course, indispensable before varnishing each part. A 45° angle cutter is also essential for those 90° corners, although professionals can do this by eye.

For the gluing we used polyurethane adhesive, with its excellent adhesive qualities for most on board surfaces. You should always sand down painted or varnished surfaces a bit before gluing, to give the adhesive a better grip.

➤ *For the finishing you also have to invest the necessary time in clearly establishing exactly what you want to achieve. If you just make it up as you go along it is going to take you forever. Once you have all your wood and equipment on board it is simply a question of cutting and working each length of moulding until it fits perfectly into place.*

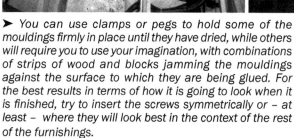

➤ *You can use clamps or pegs to hold some of the mouldings firmly in place until they have dried, while others will require you to use your imagination, with combinations of strips of wood and blocks jamming the mouldings against the surface to which they are being glued. For the best results in terms of how it is going to look when it is finished, try to insert the screws symmetrically or – at least – where they will look best in the context of the rest of the furnishings.*

Inserting screws into solid wood

Inserting screws into solid wood requires a certain operational procedure, specifically to avoid the wood splitting and to make sure the screws are firm, fully inserted and that their heads will not break off when tightened. First of all you have to drill holes of the correct diameter. The outer part (in our case the moulding) must let the screw pass freely through it (photo 1), while the hole in the inner part must have the same diameter as the body of the screw, not including its thread (photo 2). Countersinking the top end of the hole (always with the specific accessory, never using a larger drill bit) will also greatly improve the look of the finish.

foto 1

foto 2

 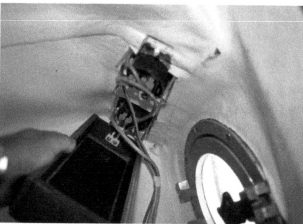 

Concealing the repeaters

➤ *On board* Samba *nothing had been planned for concealing the back of the multifunction repeaters which looked so ugly in the aft cabin. In the end the solution that we came up with was a box, which we made up from cut-offs left over from the skirting boards and teak-veneer plywood sheets. We then attached this to the inside of the hull with a couple of plastic snap-on fasteners. We also used a white spiral sheath to conceal the wiring, which now almost goes unnoticed against the white lining in the corner.*

A protector for the compressor

➤ *Installed in the storage locker, where sails and other odds and ends are continuously being taken out and put back in, the compressor for the cold box runs the evident risk of receiving knocks. To avoid this we made up this 'cage', using lengths of aluminium handrail, which can easily be bent. We also installed a shelf, with edge trim, for small objects above the compressor, fitting the whole assembly in place with two prefabricated wooden brackets.*

 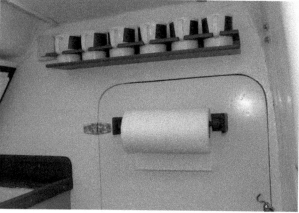

Cup holder

➤ *Two skirting board cut-offs and a length of teak (although any wood will do) served as the base for this cup holder, which we made up in just over an hour. All we had to do was make the holes in the first length of skirting board, using our hole-cutting saw, and then cut out the slots for the cup handles, sand the whole piece down a bit, apply three coats of varnish and that was that. You can buy similar accessories in teak from the chandlers, such as this paper roll holder for the galley, which you can see below the cup holder, but that does spoil the pleasure of making them yourself.*

Multi-purpose shelf

➤ A glove compartment next to the chart table to keep small objects in is an extremely practical idea and a standard fitting on many boats. We wanted one too, so we designed this one that uses the bench itself as a base with the other end supported by a batten, screwed to the bulkhead forward of the table. One way of drawing out the perimeter of an irregular shelf like this, which has to fit around wires, hoses and struts as well as adapting to the curved shape of the hull, is first to make up an approximate piece using MDF, hardboard or a piece of scrap plywood. You then provisionally fit this part into place and, using fast drying glue, attach crosscut strips of wood so that the tips follow the required outline. This piece can then be used as a template to trace out the shape on the final wooden shelf, following the points of the wooden strips. It is a process that is both simple and reliable.

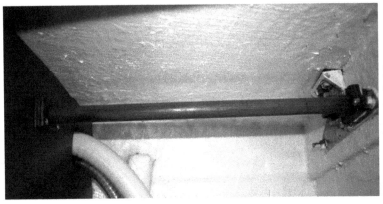

Bars in the cupboards

➤ There is no technical secret about cutting a wooden bar to length (you can buy these bars at any DIY store) and converting it into a bar to hang clothing from. The brackets that it slots into at both ends can be made up from pieces of scrap plywood by drilling holes the same diameter as the bar and then opening it up on one side. In this cupboard, where the position of the bar coincides with the internal anchorage point of the chainplate, we used polyurethane adhesive to attach the bracket on the aft side. To avoid your clothes hangers rattling about from side to side with every movement of the boat, fit a wooden clothes rail and use plastic hangers.

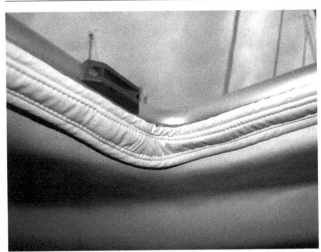

Finishing off the hatches

➤ One unexpected problem that arose as a result of the new cushioned lining on Samba's ceilings was that afterwards there was no way that we could refit the original wood mouldings around the hatches which we had been quite prepared to restore. The solution that we came up with was to go back to the upholsterer and get him to make up some broad trims, which we then glued against the inside frame of the hatches as a finish.

Making the new from the old

➤ For many years, lying around our junk room at home, there was a series of attractively grooved old wooden panels, which had previously decorated the back of my grandmother's wardrobe. Somehow, and who knows why, they never got thrown out and, while we were refitting Samba, we came up with the perfect use for them. After sanding them down and varnishing them, we installed them as back rests behind the chart table and saloon benches. I have seen similar cases where wardrobe doors or bedside tables, drawer handles, tabletop lamps or other wooden finishes have been recovered from old furnishings and given a new nautical lease of life. These ideas contribute a personal touch to any refit.

Side shelves

➤ After the cupboards that the carpenter made up were on each side of the saloon, we divided up the remaining space, previously occupied by the old berths, by installing a large shelf. Generally speaking stowage on boats is more practical when the space is divided up (panels, shelves, drawers, etc.) rather than relying on large cupboards or lockers.

Supports and attachments

➤ Among the infinite finishing details are the different systems for holding doors, covers and drawers in their open position. For the chart table and cupboard doors we decided on a spring system that is very versatile and easy to install and operate. On two of the doors we used classic hooks and eyes, installing them at the top of the door so that we would not have to bend down to work them, as you have to do on a lot of boats. For the door to the head we chose a pressure latch and for the cutlery tray, to stop it sliding open when sailing heeled over, we used a simple bolt.

Painting and varnishing

➤ *A number of furnishing parts that could not be removed had to be varnished and painted on board. When this is the case you have to take great care not to stain the rest of the interior, now in its final cosmetic stages. On dry land we also applied three coats of paint to the doors and three coats of satin-finish varnish to the mouldings and shelves. When painting it is always best to stick with one make of paint, particularly for adjoining areas. A simple white, for example, will have different tones depending on the manufacturer and a difference can be seen, such as the example of the touching up of the storage locker door (photo on the right), where you can make out a rectangle that is a different tone from the rest of the door. As this particular bit of bodging is on the inside surface of that door we decided to let it go. With varnishes the problem is not as evident but even so you must take care. In the photo on the right you can also see a section of broken Formica that we stuck back in place on the furnishing unit in the galley. When carrying out a refit not every solution will be perfect.*

Hiding away the flue hoses

The ducting that we made up to hide away the breather pipes where they pass through the saloon turned into a real engineering problem. It was impossible to design the perfect cover, one that would simultaneously allow for the opening of the side cupboard, hide the offset in the ceiling, at the corner of the roof panel, and at the same time remain discreet. The best compromise that we could come up with left the ends of the pipes in sight. When we get a chance we are going to try replacing the dorade box with a white polyamide model, and maybe then we could also use a length of white pipe for the final section. Damned details!

Fortunately human memory is a fragile thing and, shortly after we had finished our refit, although there were always going to be details that need to be sorted out, all the bad times, the anguish and desperation of the past months, as if by magic, evaporated into thin air.

Done!

Along the way there were interminable photographic sessions bathed in summer sunshine or steeped in winter's cold, endless weekends stolen from the family to eat dust on board, a constant state of anxiety caused by the lack of progress accompanied by doubts as to the final effectiveness of the solutions that we were resorting to. But now our horizon had changed. Now, at last, we could take *Samba* back to sea.

The question that you will most often be asked after your refit is, how much did it cost? A difficult question to answer, in fact it is absurd even to try. Recovering a boat from her ostracism is not an investment. It is something you do for fun or because you have a dream, because you want your boat to be special. From the financial point of view most refits are simply a non-starter, and my first recommendation for anyone who is thinking of setting out on a similar adventure is to make sure they like the boat, it's style, design, size, specifications and comfort. If not, at the end of the day, you will find yourself with the emptiness of having expended a great deal of effort and money for no good reason.

My second piece of advice would be to try not to

have your boat laid up any longer than strictly necessary. Although, looking back, I am incapable of imagining how we could have gone about *Samba*'s refit in any other way than the way we did, if I had to do it again I would try – God knows how – but I would try to establish a schedule that would allow us to have her sailing again sooner.

The best solution is probably to take one thing at a time. One year you do the deck and the next the interior furnishings, or the upholstery and linings. And in the meantime, during the summer months, you can keep on sailing.

Nautical DIY requires an enormous dedication in terms of time and one ought to be very careful when evaluating that most precious commodity of modern life. Dedicating each and every weekend the whole year round to DIY is going to represent around 400 hours, roughly the time that a professional tradesman could do in only two and a half months.

If you take into account the fact that a refit like *Samba*'s needed 1,800 man hours, then doing something similar without professional help would take you about five years even if, and it is a big 'if', as an amateur you could do the work as efficiently as a good professional, with all the experience and know-how that allows him to dive straight in, whereas you have to experiment and find your way.

Most of the refits started by enthusiasts run out of steam well before they are finished or end up being bodged jobs, if not both at the same time. The main problem is a tendency to overestimate one's own technical skills or the amount of time one is able or prepared to invest.

Apart from the profound satisfaction of actually completing a project, the main gain from a complete refit, or any partial refit or repair, is the knowledge that you acquire. Right now, off the top of my head, I could tell you, almost without hesitation, the diameter and size of each bolt and screw, nut and washer used in the refit, where all of the wires, ducts and channels run, or the best way to strip down and reassemble just about everything on board. Mind you, at the same time I could also tell you exactly where one of the panels is discreetly peeling or point out that run in the paint, or the hair from the brush that will be forever preserved in her varnish.

Refitting your boat, above all when you are involved in a more active way than just signing the cheques, requires a whole lot of tenacity. It is a bit like running a marathon. You have to find your rhythm and then get your head down and hang on in there until you have crossed the finishing line. If you set off at a sprint you will run out of steam before you get half way.

➤ *With her hull and deck brought up to date* Samba's *veteran, yet still attractive, silhouette looks so much better. Her design continues to generate the admiration of all who love a boat with character.*

➤ *Fortunately human memory is a fragile thing; as soon as* Samba *had a few sea miles under her belt, all the trials and tribulations of her refit were nothing but a distant memory.*

➤ *The pleasure of sailing* Samba *was like a balm after so much work. At last our horizon is back where it should be: out to sea.*

Step by step

➤ **Saloon:** *From an aesthetic point of view, one of the most improved aspects of Samba's renovation is her interior, through the visual sense of space and brightness achieved by the use of white surfaces as the basis of her decorative scheme.*

➤ **Linings:** *Renewing floors, ceilings, walls and upholstery had a decisive effect on the new gleam given off by her interiors.*

➤ **Galley:** *This is the part of the boat where the work was closest to being a restoration in the strict sense of the word. Apart from replacing broken parts (worktops, sinks, cooker, etc.) and making up those that were missing or had not originally been included (doors, pressurised water, cold box, etc.), the rest of the work was essentially a matter of deep cleaning and then adding a few good coats of paint.*

➤ **Chart table:** *If the change in the navigation station is radical, the most complex part of the work is hidden away behind the control panel. The original electrics and electronics have now been brought completely up to date.*

➤ **Attention to detail:** *At the end of any refit, the part that is noticed most is the attention that has been given to the details. You have to keep your standards as high as possible until you have finished the job.*

➤ **Forward cabin:** *A quick glance at these photos renders any commentary superfluous. However, we would like to say that the new furnishings were re-made based on the boat's original bulkheads and panels. Doing it any other way tends to complicate things in any refit.*

➤ **Aft cabin:** *Just by replacing the foam and upholstery, and lining the walls, the aft cabin now looks completely different.*

➤ **Head:** *A comparison of these two photos makes it difficult to believe that they are of the same part of the boat. Samba's head did not exist on the port side, with the remnants of the original berths that occupied this space still in place as shelves. No further comment is necessary on the change achieved.*

➤ **Head:** *On the starboard side we also modernised the head considerably, gaining enough space to extend the cupboards. Following her refit, Samba now has a head that would not look out of place on a 50-footer.*

➤ **Wiring installation:** *This has been one of the most laborious tasks of Samba's refit yet, paradoxically, there is barely any visible sign of it (something that could not be said about the old installation) apart from the control panel and light fittings.*

 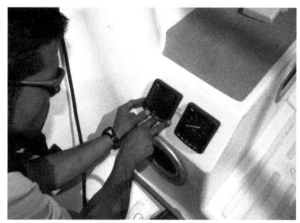

➤ **Electronics:** *These days it is hardly ever worthwhile repairing electronic instruments. Apart from the problem of finding spare parts, the specifications of the equipment itself are constantly improving in leaps and bounds, while the prices have hardly changed at all.*

➤ **Engine:** *After a certain age (20/25 years), or number of operational hours (1,500/2,000), you can safely assume that a sailing boat's engine has just about done its bit. Of course, as long as it keeps on working, you can delay actually replacing it, but you have to accept that it has reached the end of its useful life and that it is no longer worth investing in major repairs.*

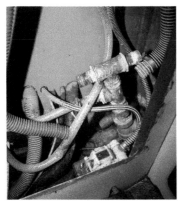

➤ **Freshwater installation:** *On board a boat, replacing pipes, hoses, pumps and through-hulls is something that, as time goes by, has to be done on a systematic basis. This work, essential in terms of safety and comfort, is also easy enough for most amateurs to get to grips with.*

➤ **Deck:** *Stripping down, fixing up and painting the deck, in order to later install new or restored deck hardware and accessories, was the job that launched* Samba's *refit and, at the same time, one of the last jobs to be finished. The results, both aesthetic and functional, are there for all to see.*

Refitting a used boat: 35 pieces of essential advice

1. **Allow for the worst, then multiply by two:** For some strange reason, the time and money that you calculated to invest in a job tends to be twice your original estimate. Always be prepared for the worst.

2. **DIY is not a way of saving money:** You have to think of Do-It-Yourself as a hobby, never as a way of actually saving money on paying a professional. In financial terms, it is always more profitable to put in overtime hours at your own place of work and use the money to pay someone else to work on your boat.

3. **Everything has to be learnt:** In the same way as a recently qualified medical student is not ready to make a liver transplant, anyone who is new to DIY must also accept that they will have technical limitations. Blithely crossing this threshold will be a source of enormous frustration and serious problems.

4. **One thing at a time:** Rather than trying to get a thousand and one things done on a Saturday morning, limit yourself to one or two and try to take them from start to finish.

5. **Planning and preparation:** Before getting down to the work on board make sure that you have everything you need. Nothing wastes time as much as having to run off to the hardware store or chandlers. For big jobs you will have to establish a sequence and try to estimate the amount of time required for each step.

6. **Use the right tools:** Having the right tool kit is half of any repair. Always use the right tools and try to buy the best. Tools are also personal things; avoid borrowing them from your neighbours or tradesmen.

7. **Time-keeping:** If you have decided to spend the weekend working on your boat try to work to a fixed timetable, avoid those endless coffee breaks or laying down tools to chat to friends. Tempus fugit.

8. **Look before you leap:** Before starting a job on board make up a visual composition of all the steps to be followed.

9. **Before drilling a hole or making a cut count to 1,000:** This is a fundamental maxim. Never drill a hole or make a cut without looking to see what lies on the other side or without checking, over and over again, that you have got the measurements just right.

10. **To fix or replace:** This is one of the great dilemmas of any restoration. Everything will depend on how fond you are of the item in question, how much replacing it is going to cost, the technical difficulties involved or the complexity of finding spare parts for old equipment. The solution is never easy.

11. **Brush, roller or spray:** Painting and varnishing are two of the jobs that amateurs most frequently feel confident about doing. While they may seem simple, getting the best results can be quite complex. If you are a beginner, start with parts of the boat that are hidden from sight (bilges, locker lids, etc.). Do not start by painting her topsides or varnishing the top of the saloon table.

12. **Seek excellence:** Do not be satisfied with a bodged job, set yourself professional goals and standards. Despite your best efforts, reality inevitably comes along and interferes, reducing the technical heights you aspire to.

13. **Simple is best:** All things being equal, in terms of the technical result or cost, always go for the simplest option. As far as possible steer clear of complicated solutions.

14. **Professional or amateur:** Combining amateur work, the stuff you can do yourself, with professional work will limit the amount of time and money that you have to spend on your refit. If you choose a well-balanced combination you can cut the costs by doing the unskilled work yourself, particularly in areas where you have more know-how (electrics, carpentry, painting) and leave the skilled jobs to professionals.

15. **No way back:** In most important repair jobs you will reach a turning point from where there is no going back. In these cases you just have to keep on going until you get to the end.

16. **Don't throw away anything that might be useful:** This piece of advice may, on occasions, help you to alleviate problems arising from the last one.

17. **Practice makes perfect:** Do not expect your first DIY efforts to produce spectacular results. Little by little your technique will improve as you learn from both your mistakes and the things you get right.

18. **Standardise materials:** Some shipyards use the same locker lids or cupboard doors for all the boats in their catalogue. You should also try to unify hose diameters, types of screws and bolts, make and colour of paint or the sizes of the wooden strips and battens used for the carpentry work.

19. **Don't be too shy to ask:** No competent tradesman will deny you a piece of good advice regarding a specific repair. You must, however, bear in mind that asking for advice is one thing and expecting a free weekend course in electronics, for example, is something quite different.

20. **If in doubt don't:** Do not take on any job on your boat unless you are convinced of its utility or result and certain about the steps that you will need to take.

21. **Be bold:** You always need to be a little daring when it comes to uncharted territory. At the same time try to be cautious and get the best advice about DIY jobs that are new to you. You will often find that these are simpler than they seem.

22. **Set yourself achievable goals:** Be demanding about your aims, and do not set yourself unachievable goals. Having a full-time job is one of the biggest frustrations for any amateur.

23. **Cheap ends up expensive:** Bodged jobs cost double. You end up dissatisfied and, within a short time, deciding to either do it or have it done again, but this time correctly.

24. **Do not invest in a boat that you do not like:** If you have to invest time and money in a boat, try to make sure that it is one that you are happy with, in every sense.

25. **Keep the lay-ups short:** Having your boat in dry dock during the sailing season is unbelievably frustrating. Try to plan the work so that you can actually get some sailing in as well, at least during the summer months.

26. **Tenacity is a virtue:** List the jobs that need to be done and draw up a schedule for doing them, one by one.

27. **Work in a neat and orderly way:** Keep the boat and your tools and equipment clean and well ordered at all times; before, during and after you have finished each job.

28. **Think ahead:** When you have taken your boat out of the water to renew the antifouling paint, for example, is a perfect opportunity to have a look at your propeller shaft, change the through-hulls or check the play in the rudder.

29. **Do not leave work half done:** Make an effort to finish any job that you start. Leaving a job half done is worse than not starting it at all.

30. **Read the instructions:** Spend as much time as you need doing this. In the long run it will save you far more time when assembling parts and accessories.

31. **Weights and measures:** A tape measure and callipers are indispensable components of your tool kit. Do not take pot luck when it comes to weights and measures.

32. **Other people's inventiveness:** Be a good observer of the solutions, of all shapes and sizes, that have been adopted in other people's boats. They can be an endless source of inspiration.

33. **Pay attention to detail:** One of the main differences between professional and amateur work is in attention paid to detail, evident both in the work itself and the way that it is finished.

34. **The chains breaks at the weakest link:** A bad connection or screw, a loose rivet or clamp can spoil the best installation.

35. **Avoid shortcuts:** Do not be complacent about the work. As soon as you have your boat back in the water any bodged jobs that you did during the refit will quickly become evident (breakages, water leaks, malfunctioning, etc.) and the only one to suffer as a result will be you.

Waste management

Installing a waste tank had already been anticipated as part of the refit although, in the clamour of the battle to solve bigger on-board problems, it somehow got left aside. As we had Samba back in dry dock we decided not to put it off any longer.

A more ecological toilet

Unlike a boat's electronics, rigging or galley, the toilet is something that each member of your crew or passengers will be using on a reasonably regular basis. From the most skilled to the clumsiest, they are also going to have to manage all on their own, practically without any help or the possibility of any operating errors. Never has the saying 'simple is best' been more true than in this case, where everything you can do to simplify head operations and waste management will help to avoid problems with or the incorrect use of the system.

Generalising about the best way to store and discharge your organic waste is a complicated business. Each boat is different, each system has its advantages and disadvantages and each owner has his own particular preferences and needs. Apart from explaining the system that we adopted for *Samba*, we would also like to take a quick look at other types of installation, along with some of their pros and cons.

There is a high level of compatibility between the different makes of toilet bowls, discharge pumps, tubes, valves and holding tanks, which makes problem-solving far easier and allows for modular assemblies that can provide an all but perfect solution to your particular problems.

Wastewater discharge in port

Part of the legislation regarding storage and discharge of wastewater is based on the assumption that ports are equipped with machines that can pump this waste out of your boat. At the time of writing, few ports actually have these facilities and we hope that, when they do, the system adopted will be based on an agreed European standard. If each marina just goes ahead and applies its own criteria, as sadly has been the case with freshwater connections and 220 V supply lines, then we are going to have a problem on our hands.

Whatever material you use and wherever you locate your holding tank, the port system must be able to connect up to the top or the bottom of the tank to discharge it. In either case this discharge may be either independent or fitted in a Y- fitting with the sea discharge line. In order to avoid having lengths of piping, a Y-fitting or having to make a hole in the deck for the port discharge connection, in some installations you can fit a suction cap directly on top of the tank.

Installation

Any owner with a bit of skill and a free weekend in front of him will be able, on his own, to adapt his waste-water system. Once you have got the necessary material on board all you have to do is join up a few pipes to a holding tank and the necessary valves until you have set up a new circuit.

The dilemmas and complications generally arise at a much earlier stage, when you have to design it, prior to installation. First you have to decide on the best position for your tank, limiting the required lengths of pipe as much as possible (thus increasing pump effectiveness) and also finding an unobstructed run for the piping, plus convenient accessible positions for valves and/or bypasses. Apart from all of that there are no particular mysteries involved.

It is a good idea to consult a number of different installation diagrams, which you can find in the leaflets provided by marine sanitation suppliers, and it is also interesting to see how this problem has been solved on other boats, either in the original design or as a refit. The prize for the best combinations of effectiveness/ simplicity/low cost goes to the large production boat builders that, generally speaking, are an excellent source of inspiration when it comes to finding the simplest, cheapest and most effective way.

➤ *The convenience (no pun intended) of having a black water tank when cruising is evident, avoiding the need to go onshore to use the loo whenever you are in port or anchored in protected waters.*

Step by step

Holding tanks: materials and location

The two most commonly used materials for waste tanks are rotomoulded polythene and stainless steel. You also come across them in fibreglass, but these are only permited when they have been fitted at origin by the shipyard. Polythene tanks, on sale prefabricated in different sizes, are lighter and more economical than stainless steel ones. On the other hand the advantage of stainless steel tanks is that you can have them precisely made to fit your locker, that space in the bilges or a cupboard, which is a plus in many cases.

The height of the tank in your boat (above or below the waterline) also affects the rest of the installation. If you fit it above the waterline, as is the case on many production boats, discharge is gravity fed, avoiding the need for a pump. However, this layout can be problematic, in terms of the added weight in the upper works and the loss of convenient stowage space. Obviously, if you fit the tank below the waterline it will be more discreet and its weight will be more centred, but then again you are going to need a discharge pump.

Holding tanks installed above the waterline discharge from the bottom, in so far as this guarantees that it will be completely emptied. However, when the tank is discharged using a pump it is better to suction the contents out from above, with the inlet located as low as possible in the tank. This avoids any organic residues that have not been discharged from stagnating in the pipes.

There are a number of appropriate, or possible, locations for installing your wastewater holding tank. One of the best options for boats with a head at the stern is the cockpit locker. In this case it is quite easy to install it even above the waterline, freeing up space at the bottom of the locker.

Things are a bit more complicated when the head is located at the centre or the bow of the boat, as you have to find a space in one of the cupboards or the bilge. It is not advisable to install a holding tank in lockers directly under the saloon benches or the forward berths. The risk of bad odours, even if they are barely detectable, is high and tends to take a little of the shine off life aboard.

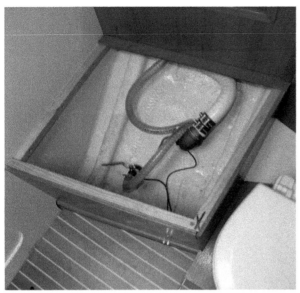

➤ In the case of Samba we took advantage of a space that had previously been occupied by the through-hulls of the original head. Here we installed a 50-litre capacity stainless steel tank. Once we had blocked off the old through-hulls, the water inlet was moved over next to the bowl. Holding tank discharge will be by suction using a combined pump and filtration unit, with an outlet just above the waterline.

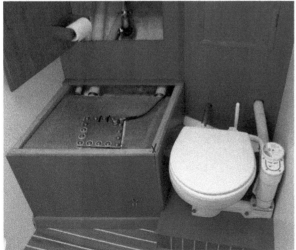

➤ One of the advantages of stainless steel tanks is that you can have them made up to any shape or size you want. This means that they can be made up to fit neatly into your lockers or bilge bottom. We made our order after establishing the dimensions with bits of scrap wood stuck together and then marking the shape out on sheets of cardboard and cutting them to size.

➤ The wastewater holding tank in this example is above the waterline. To empty it all you have to do is open the bottom valve and the pipe empties out directly into the sea. The top pipes are the inputs from the bowl, suction from the deck and the breathing tube.

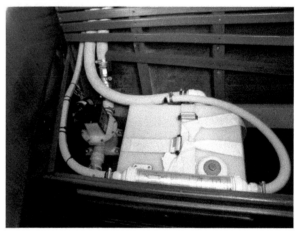

➤ As the head on this boat is located amidships, the polythene wastewater tank and its discharge pump had to be installed underneath the forward berth.

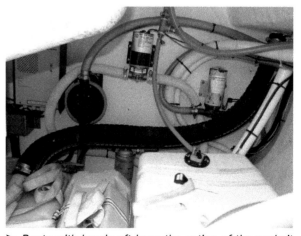

➤ Boats with heads aft have the option of the cockpit locker for installing both tank and discharge pump. This also has the advantage of distancing eventual bad odours from life on board.

Discharge pumps and maceration units

The location of the holding tank determines the kind of pump to be installed. If the tank is above the waterline you do not need a pump at all. Even if the outlet is below the waterline, the speed of the boat will create a suction effect. The faster you are going, the better your tank will empty, like the cockpit drains on small dinghies.

Now that our waste is going to be evacuated on the open sea it is not a bad idea to move the discharge outlet to a level slightly above the waterline (10/20 cm). The pump (or gravity as the case may be) will work better without the resistance of the water, while you can also avoid putting another hole in the boat's hull. Even in this case, however, a seacock is still recommended; in fact, according to the legislation, it is obligatory.

If the tank is below the waterline you can choose pumps that are either hand or electrically operated. The former are usually membrane pumps, a cheap and reliable system allowing for a high level of flow, although they are not particularly fond of discharging solid or sharp objects, something that you should not normally find in your waste tanks.

Electric wastewater pumps exist with either membrane or flexible impellers, the first merely adding electrical power to the advantages and disadvantages of their manual equivalent. Impeller pumps are quick, convenient to use and can deal with small pieces of solid waste without any problem at all. On the other hand they respond badly to running dry and you have to be attentive to switch them off as soon the tank has emptied.

➤ *To improve the effectiveness of your system, the tank, discharge pump and outlet to the sea should all be as close to each other as possible. Even though the discharge is at or above water level (as in the case of Samba) the through-hull should be fitted with a seacock.*

➤ *Replacing the manual toilet with an electric one represents a big improvement in terms of ease of use and would only take you about 15 minutes to do. The electric pump that we used is also a macerator unit (which will almost liquefy the waste), ensuring that discharge will be even more efficient.*

Valves, bypasses and pipes

One of the main sources of bad odours in toilet circuits comes from the pipes themselves. The classic white hoses with their metal spirals (a standard fitting on most boats) are slightly porous, which means that they become impregnated and eventually, over the years, inevitably make their contribution to that characteristic head smell. Modern pipes with a triple coating (much more expensive by the way) avoid this problem.

Whatever the type of pipes you decide to use, one of the goals when designing the installation is to avoid sections in which the wastewater can stagnate. There is no simple solution.

With regard to the through-hulls, bypasses, Y-valves or seacocks, these are usually made of brass, a material that is quite cheap and sufficiently resistant in a marine environment. Stainless steel, a longer lasting and harder metal, can also be used, as can black fibreglass reinforced polypropylene, which ought not to be confused with its polyamide counterpart (usually white). Valves, hoses and PVC fittings, standard in industrial plumbing, generally recognisable because of their grey colour, are not recommended for on-board use.

Level indicators are legally required. These can be electrical (possibly remote) or mechanical (fitted to the tank). Translucent polythene tanks allow you to actually see the level, as long as it is partially visible.

Electrical pumps, in the toilet bowl or extractors, tend to consume a lot of electricity. Apart from checking the capacity of your batteries (although normally these pumps would only be used in short bursts), you should also connect them up with 4, 6 or 10 mm^2 wires, depending on length, in order to avoid problems of power loss or overheating.

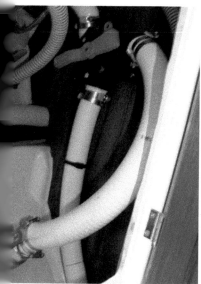

➤ Each pipe, pump, bypass or seacock installed on board is a potential source of problems. The simpler the installation the better. This consideration, apart from its affect on your budget, is the best solution for avoiding operating problems while limiting the risk of breakdowns (leaks, breaks, bad smells, ingress of water or organic waste, etc.).

➤ Teflon tape (PTFE) and silicone are some of the most commonly used materials for avoiding leaking through-hulls. Some shipyards, seeking the best solution in terms of cost effectiveness, trust the tightness of their seals to polyurethane sealant, applying it to all of the joints and threads that make up the circuit. Fittings below the waterline should be always double-clamped.

➤ There are switches with keys to avoid discharging the waste into the water when this would not be appropriate. In our case a clear warning 'DO NOT TOUCH' written on the pump switch also tends to get the message across.

➤ A breather pipe is necessary for the system to function. Eventually, an anti-smell filter will avoid organic waste odours being noted on deck, coming up through the tank's breather pipe. You should install it if you find the need to.

Steering system

We had taken a good look at Samba's rudder, stock, quadrant and tiller cables on a number of occasions and all of us, professional and amateur alike, agreed that everything looked just fine, that the rudder assembly was one of the few areas that needed no work at all.

Appearances can be misleading

Nevertheless, on our first few outings we realised that all was not as it should be with the rudder (vibrations, a juddering sensation when we turned the wheel, a number of rather dead angles, etc.). For peace of mind, as soon as we could get *Samba* back in dry dock, following the 'official' culmination of her refit, we stripped down the quadrant, its cables and the rudder stock in order to carry out a more detailed inspection. What we found is a graphic example of just how misleading appearances can be.

A problem with the rudder, even more than a propulsion problem (sail or engine), will reduce a boat's capacity to manoeuvre. You should always pay very close attention to the steering mechanisms, particularly when stripping them down and putting them back together again, especially as it is not going to cost you too much in terms of either time or money.

If the final purpose and the operational properties of a rudder are virtually identical, for whatever boat, sailing or motor, the mechanisms themselves rarely coincide in terms of shape or design. Due to a lack of time or space in this book it is not possible to make a list of the innumerable types of steering systems in existence. We hope that a brief description of how we stripped down, repaired and put *Samba*'s rudder back together again will be sufficiently illustrative.

Perhaps at this point we should also point out, in a more general sense, that, among the many problems that you may encounter with the steering, one of the most recurrent symptoms of malfunction is excessive play of the rudder blade – easy enough to spot when you have her on dry land. If there is too much play in the rudder (lateral, vertical or turning) then the odds are that, sooner or later, you are going to have a problem with your steering system.

Step by step

➤ The above photo shows the quadrant after we had stripped it down (with the clamp on the right) and the stainless steel post that joins the quadrant to the rudder stock. The welding on the post, joining the parts with a different diameter, had to be renewed, having deteriorated as a result of electrolysis. This problem is impossible to detect without stripping down the whole assembly but, if the post had broken while sailing, then the rudder would have been rendered useless and we would have been all at sea without steering.

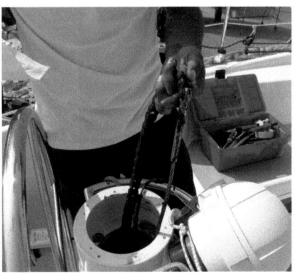

➤ To be on the safe side we also stripped down and removed the control cables, replacing them with new ones. The cables were easily removed, once we had unbolted the quadrant, and replaced them through the pedestal.

➤ Given that the quadrant was not as well preserved as we originally thought, we also decided to remove the rudder, now only held in place by the bottom pintle.

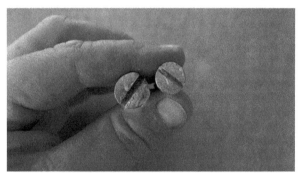

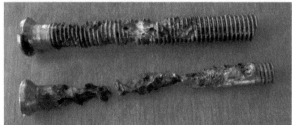

➤ Appearances can be misleading. If you look at these two bolts (top photo) that held the bottom pintle in place on the skeg, they both appear to be in perfect condition. However the reality was very different (bottom photo). Electrolysis had attacked them internally with the result that the stainless steel looked like hungry mice had been nibbling at it. The bolts in the photo are the best of the four holding the bottom pintle in place (the rest disintegrated on us). The pintle could have just dropped off at any time, with nothing to stop the rudder following it.

➤ After removing the bottom pintle we took advantage of the opportunity to recondition the plastic bushing, eliminating the lateral play at the bottom end of the rudder blade.

➤ Another reason for the excessive play of the rudder was the presence of delamination at its top end, also invisible when the rudder was in place, as it was covered over by the hull and the skeg.

➤ Before we could re-laminate the top part of the rudder with fibreglass we had to thoroughly clean up the most badly damaged existing layers of fibreglass. It was also necessary to recess the fibreglass so that the new layers laminated onto it did not become lumpy.

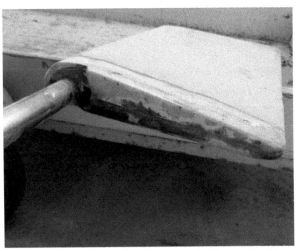

➤ The filtration of water into the top part of the rudder blade was indicated by the darkened tone of the rudder's interior plywood reinforcement. We managed to eliminate this excess damp by means of forced heat-drying.

➤ Once the blade had been dried out, we laminated several layers of fibreglass to it, until it had taken on a new consistency.

➤ With fibreglass and polyester resin it is important to reinforce the point where blade and stock meet. This is the point where most rudders suffer from water infiltration.

➤ With the rudder blade removed, and even though we did not observe any abnormal buckling or water infiltration, we also took advantage of the opportunity to reinforce the point where the skeg is attached to the hull.

➤ When we had the rudder back in place we were ready to reassemble the quadrant, but before we did that we renewed the packing in the top rudder tube, which almost certainly had not been changed for thirty years. This was another source of the rudder's excessive play.

➤ After stripping down the steering system, checking it over and putting it back together again, the only thing that remained to do was to adjust the cables running between the tiller and the quadrant. The correct tension for these lines is achieved when the wheel ceases to have any dead points, no more no less. If the cables are too loose, apart from creating an unpleasant slackness in the wheel, they can also slip out of their housings in the quadrant. If they are too tight it will be hard to turn the wheel and the whole system (sheaves, organisers, bushings, etc.) will be over stressed and may give way at some point or other.

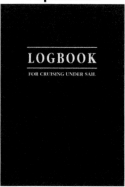
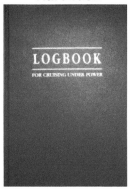

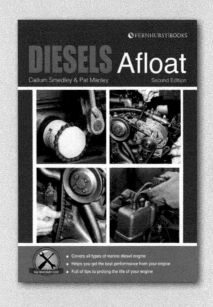
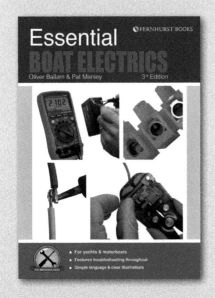

A selection of our **SEAMANSHIP** books...

The Skipper's Pocketbook
A pocket database for the busy skipper including seamanship, navigation and weather.

The New Crew's Pocketbook
Supply your crew with all the basic information they need to know on board.

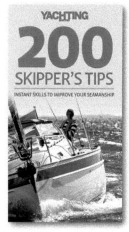

200 Skipper's Tips
Tom Cunliffe's top tips covering everything from seamanship, safety, navigation and life on board.

Sail and Rig: The Tuning Guide
The complete guide on how to trim sails & tune rigging for all conditions, making the boat sail faster & safer.

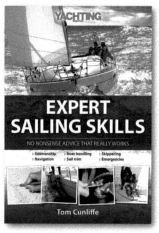

Expert Sailing Skills
A wealth of sailing & seamanship skills from the best practical articles of *Yachting Monthly.*

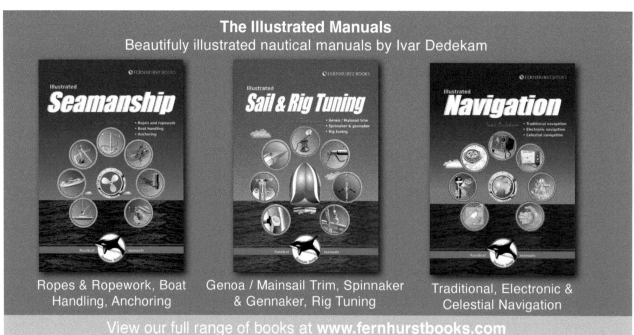

The Illustrated Manuals
Beautifuly illustrated nautical manuals by Ivar Dedekam

Ropes & Ropework, Boat Handling, Anchoring

Genoa / Mainsail Trim, Spinnaker & Gennaker, Rig Tuning

Traditional, Electronic & Celestial Navigation

View our full range of books at **www.fernhurstbooks.com**

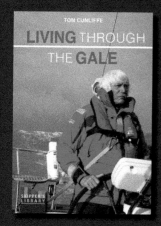

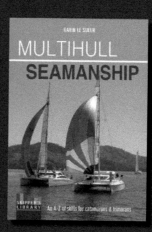
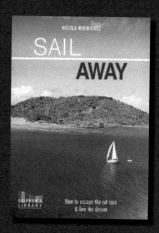

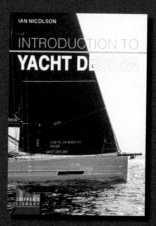

PRACTICAL COMPANIONS

Handy 24 page, on-the-water reference guides containing all the essential information for when you need it most

COCKPIT COMPANION
Basil Mosenthal

NEW CREW'S COMPANION
Basil Mosenthal

WEATHER COMPANION
Tim Bartlett

PASSAGE PLANNING
Alastair Buchan

NAVIGATION COMPANION
Tim Davison

NAUTICAL CALCULATION
Alastair Buchan

VHF COMPANION
Sara Hopkinson

ELECTRICS COMPANION
Pat Manley

DIESEL COMPANION
Pat Manley

KNOT COMPANION
Tim Davison & Steve Judkins

SPLICING COMPANION
Gareth Lincoln

FIRST AID COMPANION
Sandra Roberts

EMERGENCY COMPANION
Jon Winge

GRP REPAIR COMPANION
Pete & Penni vincent

ALL Paperback • splash-proof • spiral bound

Visit www.fernhurstbooks.com to see our full range of companions